T0222163

# Multithreading Architecture

# Synthesis Lectures on Computer Architecture

Editor
**Mark D. Hill,** *University of Wisconsin*

Synthesis Lectures on Computer Architecture publishes 50- to 100-page publications on topics pertaining to the science and art of designing, analyzing, selecting and interconnecting hardware components to create computers that meet functional, performance and cost goals. The scope will largely follow the purview of premier computer architecture conferences, such as ISCA, HPCA, MICRO, and ASPLOS.

Quantum Computing for Computer Architects, Second Edition
Tzvetan S. Metodi, Arvin I. Faruque, and Frederic T. Chong
2011

High Performance Datacenter Networks: Architectures, Algorithms, and Opportunities
Dennis Abts and John Kim
2011

Processor Microarchitecture: An Implementation Perspective
Antonio González, Fernando Latorre, and Grigorios Magklis
2010

Transactional Memory, 2nd edition
Tim Harris, James Larus, and Ravi Rajwar
2010

Computer Architecture Performance Evaluation Methods
Lieven Eeckhout
2010

Introduction to Reconfigurable Supercomputing
Marco Lanzagorta, Stephen Bique, and Robert Rosenberg
2009

On-Chip Networks
Natalie Enright Jerger and Li-Shiuan Peh
2009

The Memory System: You Can't Avoid It, You Can't Ignore It, You Can't Fake It
Bruce Jacob
2009

Fault Tolerant Computer Architecture
Daniel J. Sorin
2009

The Datacenter as a Computer: An Introduction to the Design of Warehouse-Scale Machines
Luiz André Barroso and Urs Hölzle
2009

Computer Architecture Techniques for Power-Efficiency
Stefanos Kaxiras and Margaret Martonosi
2008

Chip Multiprocessor Architecture: Techniques to Improve Throughput and Latency
Kunle Olukotun, Lance Hammond, and James Laudon
2007

Transactional Memory
James R. Larus and Ravi Rajwar
2006

Quantum Computing for Computer Architects
Tzvetan S. Metodi and Frederic T. Chong
2006

Multithreading Architecture

Mario Nemirovsky and Dean M. Tullsen

ISBN: 978-3-031-00610-4     paperback
ISBN: 978-3-031-01738-4     ebook

DOI 10.1007/978-3-031-01738-4

*A Publication in the Springer series*
*SYNTHESIS LECTURES ON ADVANCES IN AUTOMOTIVE TECHNOLOGY*

Lecture #21
Series Editor: Mark D. Hill, *University of Wisconsin*
Series ISSN
Synthesis Lectures on Computer Architecture
Print 1935-3235    Electronic 1935-3243

# Multithreading Architecture

Mario Nemirovsky
ICREA Research Professor at the Barcelona Supercomputer Center

Dean M. Tullsen
Professor, University of California, San Diego

*SYNTHESIS LECTURES ON COMPUTER ARCHITECTURE #21*

# ABSTRACT

Multithreaded architectures now appear across the entire range of computing devices, from the highest-performing general purpose devices to low-end embedded processors. Multithreading enables a processor core to more effectively utilize its computational resources, as a stall in one thread need not cause execution resources to be idle. This enables the computer architect to maximize performance within area constraints, power constraints, or energy constraints. However, the architectural options for the processor designer or architect looking to implement multithreading are quite extensive and varied, as evidenced not only by the research literature but also by the variety of commercial implementations.

This book introduces the basic concepts of multithreading, describes the a number of models of multithreading, and then develops the three classic models (coarse-grain, fine-grain, and simultaneous multithreading) in greater detail. It describes a wide variety of architectural and software design tradeoffs, as well as opportunities specific to multithreading architectures. Finally, it details a number of important commercial and academic hardware implementations of multithreading.

# KEYWORDS

multithreading

# Contents

# Preface

In the last 10–15 years, multithreaded architectures have gone from being exotic and rare to being pervasive. Despite the now common nature of multithreading, there are several reasons why a book on multithreading architecture is still necessary.

First, the multithreaded processors themselves are quite varied, from low-end embedded processors to the most aggressive general-purpose speed demons. These architectures may implement multithreading for wildly different reasons, from minimizing energy consumed to extracting every ounce of performance possible from a given architecture. The best way to implement multithreading will vary widely across these designs.

Second, a wide variety of fundamental models of multithreading exist and all, or nearly all, are still relevant today. Depending on the goals of the design, simultaneous multithreading, fine-grain multithreading, coarse-grain multithreading, and conjoined cores all could be viable options under the right constraints.

Third, multithreading architectures continue to be a highly active research area and the field continues to change and adapt.

Fourth, despite the fact that multithreading is now common and touches every aspect of machine and pipeline design, most current undergraduate and graduate texts only devote a small number of pages to the topic.

Chapter 2 of this book establishes the various models of multithreading, including the principle three (coarse-grain, fine-grain, and simultaneous multithreading), as well as a number of other multithreaded models less well known or not always listed under the umbrella of multithreading. Chapters 3-5 develop the three primary models of multithreading more extensively, including detailed descriptions on how to implement them on top of existing pipelines. Chapter 6 describes a wide array of research that seeks to manage contention for shared resources and Chapter 7 examines new architectural opportunities enabled by those shared resources. Chapter 8 highlights some of the challenges of simulating and measuring multithreaded performance. Chapter 9 then details a number of important real implementations of multithreading.

This book seeks to be accessible at a number of levels. Most of the book will be easily followed by someone who has been exposed to an undergraduate computer architecture class. For example, Chapters 3-5 detail the addition of multithreading on top of pipeline designs that will be familiar to someone who has taken such a class. Chapters 6-8, on the other hand, should be useful to any advanced graduate student, researcher, or processor designer/architect who wants to catch up on the latest multithreading research.

The authors would like to thank a number of people who influenced the content of this book. We thank our editor Mark Hill and publisher Mike Morgan for their invitation to write the book, their direction and advice, and especially their patience. We would like to thank Burton Smith and Jim Smith for significant conversations early in the process. James Laudon and Mark Hill provided extensive feedback and suggestions. Michael Taylor and Simha Sethumadhavan provided specific suggestions on content. Several others helped proofread and provide further content suggestions, including Leo Porter, John Seng, Manish Arora, Vasileios Kontorinis, Eric Tune, Kathryn Tullsen, Vasilis Karakostas, and Damian Rocca. We would also like to thank our loving wives Nancy Tullsen and Laura Nemirovsky for their continuous help and support. A special thanks goes to Mario's son, Daniel Nemirovsky, who motivated him on this endeavor and for his invaluable help in editing and proofreading his writing.

We would especially like to thank our many co-authors, co-workers, and co-designers over the years, so many of whom increased our knowledge and understanding of these topics, and thereby influenced this book in significant ways.

Mario Nemirovsky and Dean M. Tullsen
December 2012

# CHAPTER 1

# Introduction

A multithreaded architecture is one in which a single processor has the ability to follow multiple streams of execution without the aid of software context switches. If a conventional processor (Figure 1.1(a)) wants to stop executing instructions from one thread and begin executing instructions from another thread, it requires software to dump the state of the running thread into memory, select another thread, and then load the state of that thread into the processor. That would typically require many thousands of cycles, particularly if the operating system is invoked. A multithreaded architecture (Figure 1.1(b)), on the other hand, can access the state of multiple threads in, or near, the processor core. This allows the multithreaded architecture to quickly switch between threads, and potentially utilize the processor resources more efficiently and effectively.

In order to achieve this, a multithreaded architecture must be able to store the state of multiple threads in hardware—we refer to this storage as *hardware contexts*, where the number of hardware contexts supported defines the level of multithreading (the number of threads that can share the processor without software intervention). The state of a thread is primarily composed of the program counter (PC), the contents of general purpose registers, and special purpose and program status registers. It does not include memory (because that remains in place), or dynamic state that can be rebuilt or retained between thread invocations (branch predictor, cache, or TLB contents).

Hardware multithreading is beneficial when there is a mismatch between the hardware support for instruction level parallelism (ILP) and the level of ILP in an executing thread. More generally, if we want to also include scalar machines, it addresses the gap between the peak hardware bandwidth and the achieved software throughput of one thread. Multithreading addresses this gap because it allows multiple threads to share the processor, making execution resources that a single thread would not use available to other threads. In this way, insufficient instruction level parallelism is supplemented by exploiting thread level parallelism (TLP). Just as a moving company strives to never send a moving truck cross country with a partial load, but instead groups multiple households onto the same truck, a processor that is not being fully utilized by the software represents a wasted resource and an opportunity cost.

Hardware multithreading virtualizes the processor, because it makes a processor that might otherwise look little different than a traditional single core appear to software as a multiprocessor. When we add a second hardware context to a processor, a thread running on this context appears to be running on a virtual core that has the hardware capabilities of the original core minus those resources being used by the first thread, and vice versa. Thus, by constructing additional virtual cores out of otherwise unused resources, we can achieve the performance of a multiple-core processor at a fraction of the area and implementation cost. But this approach also creates challenges, as the

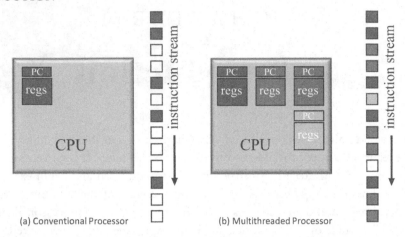

**Figure 1.1:** A conventional processor compared with a multithreaded processor.

resources available to each virtual core are far more dynamic than in a single-thread, conventional processor, as they depend on the cycle-by-cycle behavior of co-resident threads.

Because multithreading ultimately seeks to bridge the gap between hardware parallelism and software parallelism, the history of multithreading is closely tied to the advances that have increased that gap. Early computers were relatively well balanced, and did not experience large utilization gaps, except when waiting for I/O or long-term storage (e.g., disk). Those latencies were large enough that they could be hidden by software context switches (software multithreading) once time-sharing was introduced.

As computational logic speeds advanced more quickly than memory speeds, we began to see processors idle for short periods waiting for memory. This gap was bridged significantly by caches, but some memory accesses remained. Computational diversity (the need to support both simple and complex operations) also contributed to this gap. To maximize throughput, you might set your cycle time around an integer add operation, and therefore wait many cycles when executing a floating point multiply.

Microprogrammed machines were rarely idle, as the core churned through the microcode even for long latency operations; however, the introduction of pipelined processors meant that even a scalar processor could stall (incur bubbles or empty pipeline stages) on short-latency dependent operations. Superscalar processors exacerbated the problem, as even single-cycle latencies could introduce bubbles, preventing the processor from fully exploiting the width of the pipeline each cycle.

This book explores several models of multithreading. These models differ, in part, by which of these sources of the hardware/software gap they can tolerate or hide. We say a machine "hides" or "tolerates" a particular type of latency if the machine can continue to do productive work even while experiencing such a latency-causing event. *Coarse-grain multithreaded* processors directly execute one thread at a time, but can switch contexts quickly, in a matter of a few cycles. This allows them to switch to the execution of a new thread to hide long latencies (such as memory accesses), but they are less effective at hiding short latencies. *Fine-grain multithreaded* processors can context switch every cycle with no delay. This allows them to hide even short latencies by interleaving instructions from different threads while one thread is stalled. However, this processor cannot hide single-cycle latencies. A *simultaneous multithreaded* processor can issue instructions from multiple threads in the same cycle, allowing it to fill out the full issue width of the processor, even when one thread does not have sufficient ILP to use the entire issue bandwidth.

Modern architectures experience many sources of latency. These include data cache misses to nearby caches (short latency) or all the way to memory (long latency), instruction cache misses, instruction dependencies both short (integer add) and long (floating point divide), branch mispredictions, TLB misses, communication delays between processors, etc. Architects spend a significant amount of effort trying to either reduce or hide each of these sources of latency. The beauty of multithreading is that it is a *general latency tolerant* solution. It provides a single solution to all of these sources of latency. The only catch is that it requires the existence of thread level parallelism. Thus, given sufficient TLP, multithreading can hide virtually any source of latency in the processor.

This book describes the design and architecture of multithreaded processors, both those proposed in research and those implemented in commercial systems. A number of multithreaded execution models are described in Chapter 2, while the following three chapters detail the primary models of multithreading – coarse-grain multithreading, fine-grain multithreading, and simultaneous multithreading. Chapter 6 examines various sources of contention between threads and efforts to manage them. Chapter 7 describes new opportunities enabled by multithreaded architectures. Chapter 8 examines the challenge of accurately measuring and modeling multithreaded systems. Lastly, Chapter 9 describes a number of implementations of multithreading, primarily designs from industry and a few academic machines.

CHAPTER 2

# Multithreaded Execution Models

Our definition of a multithreaded processor is one that has the ability to follow multiple instruction streams without software intervention. In practice, then, this includes any machine that stores multiple program counters in hardware within the processor (i.e., on chip, for microprocessor-era machines). This definition spans the range from simultaneous multithreading to chip multiprocessing. The latter is often viewed as an alternative to multithreaded processing, but in fact is just one extreme edge of the continuum of design choices for multithreaded processors.

The focus of this chapter is on the execution models rather than specific implementations—the details of actual hardware implementations are discussed in Chapter 9, and some representative implementations are outlined in detail in Chapters 3, 4, and 5. To create as clean a distinction as possible, we focus mostly on a single region of the pipeline—instruction issue to the execution units. A typical multithreaded processor, for example, might feature simultaneous multithreading at instruction issue, but look more like a fine-grain multithreaded processor in the fetch unit.

Before beginning, we'll introduce some definitions (borrowed from Tullsen et al. [1995]) that will help to clarify the differences between the various multithreading models. With the help of these definitions and some background context, we'll see how multithreading techniques are worked into the processor to eliminate inefficiencies in fetch, execution, and other stages. The inefficiencies arise from the processor's inability to make use of execution resources when the single executing thread is stalled or blocked for some reason. Those inefficiencies, on a superscalar processor, can be classified into two categories: horizontal waste and vertical waste (Figure 2.1). A superscalar processor core is one which has the ability to sustain the execution of multiple instructions per cycle. We again focus on instruction issue to describe these concepts, but it can be useful to think about vertical waste and horizontal waste for other parts of the pipeline, as well.

Vertical waste occurs when the processor is completely unable to make use of the issue bandwidth of the processor for one cycle or a sequence of cycles. Vertical waste is typically the result of a long-latency event that completely blocks execution. The most common instructions in modern processors have single-cycle latencies, and those instructions alone cannot induce vertical waste. Multiple-cycle instructions, such as floating point operations, can induce vertical waste if there is insufficient instruction level parallelism to fill the gaps between instructions. More often, though, vertical waste is caused by the longer latencies incurred on cache misses, TLB misses, or possibly page faults. Even in an out-of-order machine, a long latency operation causes the instruction queue

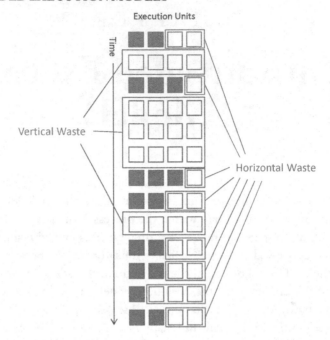

**Figure 2.1:** Vertical waste and horizontal waste of the issue bandwidth of a four-wide superscalar processor.

to fill up with dependent instructions (unless the reorder buffer fills first), preventing further fetch of potential independent instructions. An in-order processor becomes exposed to vertical waste even more quickly – the in-order processor incurs vertical waste when a single instruction is stalled for multiple cycles, while an out-of-order processor only incurs vertical waste when all instructions in the instruction queue are stalled. In both cases a branch mispredict, once resolved, can also incur vertical waste while the pipeline waits to refill with instructions from the correct branch target.

Horizontal waste, on the other hand, occurs when instruction dependencies do not allow the processor to fully utilize the issue bandwidth of the processor. They occur because on a superscalar machine even single-cycle dependencies can retard throughput. In the absence of long-latency events, horizontal waste is symptomatic of software instruction level parallelism that falls short of hardware instruction level parallelism.

A scalar processor (a processor which executes no more than a single instruction per cycle) can only experience vertical waste, but not horizontal waste.

In this chapter, we describe a variety of multithreading execution models which vary in which resources are shared between threads, how those resources are shared, and how aggressively instructions are mixed. We will move from fewest shared resources and least aggressively mixed toward more shared resources and more aggressive instruction mixing.

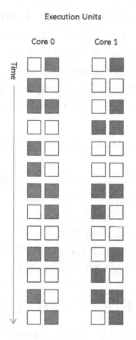

**Figure 2.2:** A *Chip Multiprocessor* can execute instructions from multiple threads at once (on distinct cores), but no execution resources are shared between threads. In this figure, and the following figures, different colors represent instructions from distinct software threads.

As mentioned, a multithreading processor stores the state of multiple contexts in hardware, and the largest part of that state is the register file. Some multithreading models require all registers to be available every cycle, some can have that storage off core. Some require a unified structure, some allow the registers to be partitioned. This can be a critical concern for computer architects, since the location and structure of the register storage will affect both the cost and performance of the multithreaded solution. Thus, we will comment on the implications of each multithreading model for the implementation of the register file.

## 2.1 CHIP MULTIPROCESSORS

Chip multiprocessors (CMPs) [Olukotun et al., 1996], also known as multicore processors, represent one end of the multithreading continuum. In a chip multiprocessor, no core resources are shared (not even L1 caches)—at most, the L2 or L3 cache is shared along with the interconnect, memory controllers, and other off-core resources. However, a CMP does have the ability to execute instructions from multiple threads in the same cycle, since execution resources are distributed among cores running different threads.

A chip multiprocessor has no ability to tolerate vertical waste or horizontal waste on a single core (Figure 2.2). However, both are reduced because the issue bandwidth is statically partitioned across cores. As a result, the losses due to either type of waste are more limited. For example, on a 4-core CMP of 2-issue superscalar cores, the total issue bandwidth is eight, yet when one thread is stalled the maximum vertical waste is only 2 instructions per cycle.

Thus, a chip multiprocessor is a multithreaded processor where all resources are statically partitioned. As a result, no thread can utilize resources in another core that the thread (if any) executing in that core is not using. However, this static partitioning does provide greater isolation and performance predictability.

In a chip multiprocessor, the register state is distributed across the cores; thus, each register file is no more complex than the register file in a single-core processor.

Chip multiprocessors and their architecture are covered extensively in other literature, and will not be discussed at length in this book. Please see the Synthesis Lectures book by Olukotun et al. [2007] for more detail. Examples of chip multiprocessors include the Intel Core Duo [Gochman et al., 2006] and the AMD Phenom II [AMD].

## 2.2   CONJOINED CORE ARCHITECTURES

There are actually a continuum of architectural options between CMPs and traditional multithreaded architectures, where some subset of core resources are shared between cores. The term *conjoined cores* [Kumar et al., 2004] has been given to a set of architectural alternatives in that space where some, but not all central execution resources, are shared between cores.

The move by industry to chip multiprocessing means that architects no longer need to respect traditional boundaries between processor cores, and the architect is free to make those boundaries fuzzy. Thus, we can maximize the utilization and efficiency of each unit by sharing pieces that a single core might not use heavily. The resources with the highest potential for sharing would be those along the periphery of the core, because they can often be shared with minimal additional latency. These peripheral resources might include the L1 and L2 "private" caches, the floating point unit, the fetch unit, and the port to the on-chip interconnect.

Depending on which resources are shared, a conjoined core architecture may look no different than a CMP with regards to how it attacks horizontal and vertical waste (Figure 2.3). For a conjoined core processor that shares the floating point unit but not the integer pipeline, for example, the floating point unit can address vertical waste, but the integer unit cannot hide vertical or horizontal waste within the core.

It is unlikely that a conjoined core processor would share the register file. Thus, the register file in a conjoined core would be the same as the register file in a chip multiprocessor and therefore no different than a conventional register file.

Examples of conjoined core architectures include the Sun Niagara T2 [Shah et al., 2007] and the AMD Bulldozer [Butler et al., 2011].

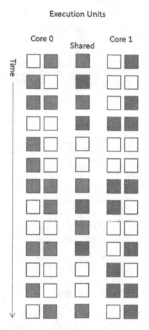

**Figure 2.3:** A *Conjoined Core* architecture only shares limited execution resources. This illustration shows an example that could represent an architecture where the integer pipelines are distinct, but the floating point pipeline is shared between cores.

## 2.3 COARSE-GRAIN MULTITHREADING

Coarse-grain multithreading [Agarwal et al., 1995, Lampson and Pier, 1980, Weber and Gupta, 1989] (CGMT), also called block multithreading or switch-on-event multithreading, has multiple hardware contexts associated with each processor core. A hardware context is the program counter, register file, and other data required to enable a software thread to execute on a core. However, only one hardware context has access to the pipeline at a time. As a result, we can only use instructions in thread B to hide latencies in thread A after thread A's context is switched out and thread B's context is switched in.

A coarse-grain multithreaded processor operates similarly to a software time-shared system, but with hardware support for fast context switch, allowing it to switch within a small number of cycles (e.g., less than 10) rather than thousands or tens of thousands. Thus, while a software multithreaded system can only hide extremely long latencies (e.g., disk), a coarse-grain multithreaded processor can hide architectural latencies, such as long cache misses or blocked synchronization (e.g., waiting at a barrier). However, it typically cannot hide short inter-instruction latencies.

Coarse-grain multithreading, then, cannot address horizontal waste, or even short stretches of vertical waste (Figure 2.4). However, assuming there is always another thread to execute, it places

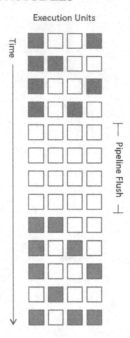

**Figure 2.4:** A *Coarse-Grain Multithreaded* architecture has the ability to switch between threads with only a short hardware context switch latency.

an upper bound on vertical waste—namely, the number of cycles it takes to switch contexts and refill the pipeline.

In a coarse-grain processor, only the currently executing register state need be available to the pipeline. Thus, the register state of non-active threads can be more distant, even off core, accessible in a few cycles. The pipeline's register file could be similar to a conventional register file, but with the added ability to save and restore the register file contents quickly. Alternatively, it could be designed as a large, partitioned register file, but never needing access to more than a single partition at once; in this implementation it is very similar to the register windows found in the Berkeley RISC [Patterson and Sequin, 1981] and Sun/Oracle SPARC processors.

The Sparcle processor [Agarwal et al., 1993] from the experimental MIT Alewife machine and the Sun MAJC processor [Tremblay et al., 2000] are examples of coarse-grain multithreaded processors.

## 2.4    FINE-GRAIN MULTITHREADING

Fine-grain multithreading (FGMT), also called interleaved multithreading, also has multiple hardware contexts associated with each core, but can switch between them with no additional delay. As a result, it can execute an instruction or instructions from a different thread each cycle. Unlike

Execution Units

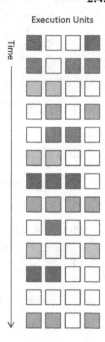

**Figure 2.5:** A *Fine-Grain Multithreaded* architecture shares the pipeline between multiple threads, with the ability to switch between contexts (threads) as frequently as every cycle with no switch delay.

coarse-grain multithreading, then, a fine-grain multithreaded processor has instructions from different threads active in the processor at once, within different pipeline stages. But within a single pipeline stage (or given our particular definitions, within the issue stage) there is only one thread represented.

A fine-grain multithreaded processor still does not address horizontal waste in a superscalar architecture, but given enough threads, can completely eliminate vertical waste (Figure 2.5). However, it also converts some of the vertical waste into additional horizontal waste, as it may still have difficulty filling the full issue bandwidth of the processor each cycle.

Prior to the arrival of superscalar processors, fine-grain multithreading on a scalar processor represented the most aggressive model of multithreading possible. However, on a superscalar processor with the ability to issue multiple instructions to the execution units each cycle, the restriction that all of those instructions must come from a single thread becomes a limiting factor.

A fine-grain machine only needs to read a single thread's registers per cycle, but may need access to a new register context the very next cycle. Thus, all registers must be close and available to the pipeline quickly. In theory, they do not need to be simultaneously readable. The barrel shifter approach of the CDC 6600 Peripheral Processors [Thornton, 1970], for example, which moved a thread's state to the processor every tenth cycle, exploits this property of only needing to make a

**Figure 2.6:** A *Simultaneous Multithreaded* Architecture has the ability to issue instructions from multiple contexts (threads) in the same cycle.

single context's registers available to the execution units each cycle. However, in modern pipelined design it is more difficult to exploit this property of the register file, because registers are read and written in different pipeline stages, which are likely to contain instructions from different threads.

The Cray/Tera MTA processor [Alverson et al., 1990] and the Sun Niagara T1 [Kongetira et al., 2005] and T2 [Shah et al., 2007] are all examples of fine-grain multi-threaded processors.

## 2.5    SIMULTANEOUS MULTITHREADING

Simultaneous multithreading [Gulati and Bagherzadeh, 1996, Hirata et al., 1992, Keckler and Dally, 1992, Tullsen et al., 1995, Yamamoto and Nemirovsky, 1995] (SMT) also has multiple hardware contexts associated with each core. In a simultaneous multithreaded processor, however, instructions from multiple threads are available to issue on any cycle. Thus, all hardware contexts, and in particular all register files, must be accessible to the pipeline and its execution resources.

As a result, the concept of a context switch between co-resident threads, which happens in a small number of cycles with coarse-grain multithreading and in a single cycle with fine-grain multithreading, is completely abandoned with simultaneous multithreading.

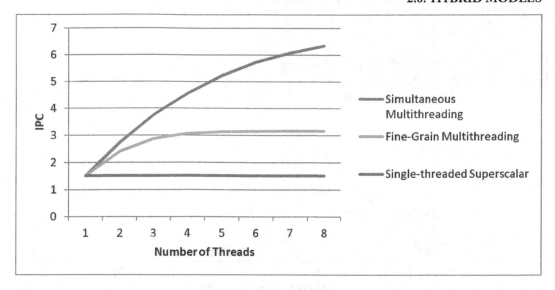

**Figure 2.7:** The performance of fine-grain and simultaneous multithreading models on a wide superscalar processor, as the number of threads increases. Reprinted from Tullsen et al. [1995].

An SMT processor addresses both vertical waste and horizontal waste (Figure 2.6). It addresses vertical waste because all active threads are able to utilize the available execution resources when one thread experiences a long latency. However, it also addresses horizontal waste because all threads are searched for instructions ready for issue each cycle. Thus, resources one thread cannot use, even for a cycle, are automatically available to other threads.

Figure 2.7 presents the performance of three processor types on a wide superscalar processor. It shows that a fine-grain multithreaded processor is ultimately limited by the amount of instruction level parallelism one thread can find in a single cycle, even when long latencies are hidden effectively. Simultaneous multithreading has no such inherent limitation.

Examples of simultaneous multithreaded processors include the Alpha AXP 21464 (announced but never released) [Diefendorff, 1999], the Intel Pentium 4 [Hinton et al., 2001], the IBM Power5 [Kalla et al., 2004], and the Intel Nehalem i7 [Dixon et al., 2010].

## 2.6 HYBRID MODELS

The models of multithreaded execution described in the prior sections are not mutually exclusive. In fact, these models can actually be highly synergistic – in modern architectures, it is common to employ a combination of these models rather than just a single model.

Sun coined the term Chip Multithreading to describe the combination of chip multiprocessing and on-core multithreading [Spracklen and Abraham, 2005]. Among existing high-performance general-purpose processors, Intel, Sun, and IBM have employed a combination of chip multipro-

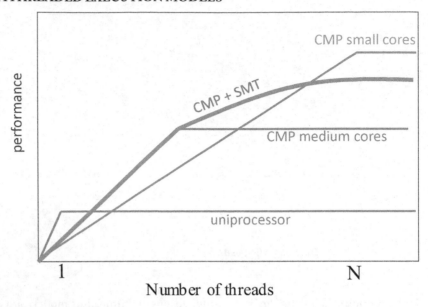

**Figure 2.8:** Chip multiprocessors can only be optimized for a single level of thread parallelism. An architecture that combines chip multiprocessing and hardware multithreading provides high performance over a much wider range of thread counts. Such an architecture, then, is much more tolerant of changes in thread parallelism.

cessing and simultaneous multithreading, Sun has employed a combination of CMP and fine-grain multithreading, and AMD and Sun have each combined CMP with conjoined cores.

Figure 2.8 demonstrates why chip multiprocessing and multithreading can be an effective combination. On a general-purpose processor, it is impossible to predict how many threads are going to be running on the processor. Thus, in the era of multicores, we have a new dimension of general-purpose processing – not only must a general-purpose processor run well across a wide variety of applications, but it also needs to run well across a wide range of thread parallelism. However, a chip multiprocessor really only runs efficiently at a single point, when the number of threads is equal to the number of cores. With fewer threads, resources are not fully utilized; with more threads, the machine fails to exploit the additional threads for improved throughput. The CMP/SMT combination, however, runs efficiently over a much wider range of thread-level parallelism, making the architecture far more robust to changes in thread count.

In proposing Balanced Multithreading, Tune et al. [2004] demonstrate that coarse-grain multithreading and simultaneous multithreading can also be synergistic. This is because processors experience latencies that are relatively bimodal—many short latencies resulting from arithmetic instruction dependencies, and relatively long latencies resulting from cache misses. SMT is highly effective at hiding the former, but can be overkill for the latter.

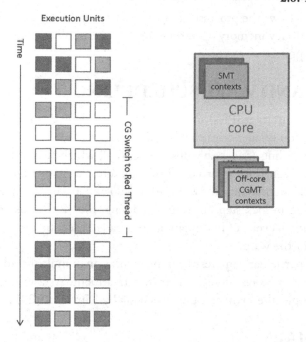

**Figure 2.9:** *Balanced Multithreading* combines simultaneous multithreading with coarse-grain multithreading. This minimizes disruption to the core pipeline by only supporting enough SMT contexts to hide short latencies. Additional contexts are stored off core, and swapped in via coarse-grain mechanisms when one of the SMT threads encounters a long-latency operation.

By combining SMT and CGMT (Figure 2.9), a balanced multithreaded processor can add modest complexity to the core (e.g., two-thread SMT), but achieve the performance of a more aggressive SMT implementation. This architecture, for example, might have two hardware contexts on the core, coupled with the ability to do a coarse-grain context swap when one of those two threads is stalled on a long cache miss. At that point, it would bring in a new context from nearby (but outside the core) thread context storage. The SMT capability in this architecture serves two purposes—it allows the processor to tolerate short inter-instruction latencies, but also allows the processor to hide the overhead of the coarse-grain context switch.

The Shared Thread Multiprocessor [Brown and Tullsen, 2008] takes the balanced multithreading approach a step further, adding CMP to the mix and again blurring somewhat the boundaries between cores. Because the thread context storage (from Balanced Multithreading, described above) is external to the core, we can now have multiple cores share it. Thus, a thread can be swapped out of one core and quickly swapped into another core. In this way the shared thread multiprocessor does for cross-core thread migrations what multithreading did for within-core context switches. The processor can now move threads to balance load, and do it orders of magnitude more quickly than the OS. This allows the processor to maneuver around events that the operating system cannot, such as long latency memory operations, or short-term phase changes that change the optimal thread-to-core mappings.

## 2.7 GPUS AND WARP SCHEDULING

Graphics processing units (GPUs), increasingly being adapted for general-purpose computation, employ a combination of single-instruction-multiple-data (SIMD) [Flynn, 1972] computing with fine-grain multithreading (and heavy doses of CMP), which they call *warp scheduling*. A typical configuration would schedule a warp of 32 threads across a smaller number (e.g., eight) cores. In a conventional SIMD architecture, those threads would each occupy a single core, with each executing the same instruction in lock step. In warp scheduling, the 32 threads are time-multiplexed across the 8 cores, requiring 4 cycles of fine-grain multithreaded execution to complete a single instruction for all 32 threads of the warp.

GPUs are a particularly interesting implementation of multithreading, but are heavily covered in other literature and indeed deserve their own treatment. As a result, we will not discuss them here. See, for example, the Synthesis Lectures book by Kim et al. [2012].

## 2.8 SUMMARY

Multithreaded architectures span the distance between chip multiprocessors (multiple program counters and register files, no execution resources shared) to simultaneous multithreading (multiple program counters and register files, nearly all execution resources shared, even on a per-cycle basis). Several of these models are synergistic, and most recent high-performance general-purpose processors employ a hierarchy of these execution models.

In this book we will focus on the three traditional models of multithreading—coarse-grain multithreading, fine-grain multithreading, and simultaneous multithreading—and the architectural issues that relate to them. We do this because CMP rightly deserves separate treatment, and the application of most of the concepts we discuss to the hybrid models will in general be straightforward to discern. We will address these three execution models in order from least aggressive to most aggressive in the next three chapters.

CHAPTER 3

# Coarse-Grain Multithreading

Coarse-Grain Multithreading (CGMT) describes a processor architecture capable of executing a single thread at a time, yet able to fully switch to execute a different thread without software intervention. In order to do so, it must have the state of multiple threads stored in or near the processor core in hardware (that state being called a *hardware context*). A context switch between two threads, each stored in a hardware context, involves flushing the state of the running thread and loading the state of another thread into the pipeline. That typically requires some small number of cycles, orders of magnitude faster than a software context switch. While a software context switch can only be used to hide extremely long latencies (e.g., disk and other I/O), a coarse-grain multithreaded processor can hide more moderate latencies, such as a cache miss that requires a main memory access or possibly a last level cache access.

## 3.1 HISTORICAL CONTEXT

As described in Chapter 1, multithreading architectures bridge the gap between available hardware throughput and available (per-thread) software throughput. As we review the origins and development of CGMT, keep in mind that it is a solution that becomes attractive when that gap is primarily the result of long-latency events.

Perhaps more than other multithreading models, CGMT has its roots in the original time-shared systems. Time sharing was introduced by Bob Bemer [Bemer, 1957] to hide the original long-latency operations (I/O), as well as to allow multiple batch-type jobs to share the processor. In those systems, software moves the state of the executing thread out of the processor, selects another for execution, and moves the state of that thread into the processor—the same operations performed in hardware by a CGMT architecture. It is worth noting that the first multithreaded systems, the DYSEAC [Leiner, 1954] and the TX-2 [Forgie, 1957], preceded software time-sharing and used hardware multithreading to hide I/O latencies that could not yet be hidden by software. Because time-shared (software multithreaded) systems preceded most hardware multithreading systems, many of the hard issues (e.g., memory sharing, protection, and isolation) were already addressed, paving the way for the later hardware multithreaded processors.

The Xerox Dorado computer [Lampson and Pier, 1980] was introduced in the early 1970s. Dorado allowed up to 16 tasks to be active in the hardware at any time; that is, the machine supported 16 hardware contexts and transitioning between tasks was done in hardware with no software intervention. In 1976 Zilog introduced the Z80 [Zilog, 1978] which included a replicated set of registers to support quick context swapping for interrupts and exceptions, enabling a limited form of CGMT between the user code and kernel interrupt handlers.

In the mid 1980s, the CHoPP corporation designed a supercomputer based on a proprietary multiprocessor. This architecture had up to 16 CPUs, each capable of executing 9 instructions per cycle, and with support for 64 hardware contexts. The hardware could swap contexts in only 3 cycles.

In the 1990s, SUN, LSI Logic, and MIT co-designed the SPARC based Sparcle processor [Agarwal et al., 1993]. This was the processor at the heart of the Alewife research project [Agarwal et al., 1995], a large, scalable multiprocessor system which needed to tolerate potentially long memory latencies. Sparcle made surprisingly few changes to the SPARC architecture to support four hardware contexts because it leveraged the register windows of the SPARC ISA [Weaver and Germond, 1994].

The MAJC (Microprocessor Architecture for Java Computing) [Tremblay et al., 2000], was also designed by Sun Microsystems as a multicore, multithreaded processor in the mid-to-late 1990s. It had 4 hardware contexts and fast context swaps. It prefetched data and instructions for stalled or non-active threads so that they would execute without delay when the thread became active again. These machines are described in more detail in Chapter 9.

## 3.2   A REFERENCE IMPLEMENTATION OF CGMT

We will be using the MIPS Instruction Set Architecture (ISA) and the MIPS R3000 specification as a baseline architecture for our discussion of coarse-grain multithreading. We will start with a detailed description of the base architecture and pipeline implementation, then describe the changes to that architecture necessary to implement CGMT. We will use the same base architecture in the next chapter on fine-grain multithreading, but a different MIPS pipeline as a starting point for simultaneous multithreading in Chapter 5.

The MIPS R3000 [Kane and Heinrich, 1992] is a single-threaded, in-order, scalar architecture with a 5-stage pipeline, as shown in Figure 3.1. Each stage is also divided into two different phases (ø1 and ø2). For instance, in the IF (Instruction Fetch) stage, an instruction's virtual address is translated to a physical address (through the instruction TLB, or ITLB) in phase ø1 and in ø2 the physical address is sent to the instruction cache to retrieve the corresponding instruction. During the Read (RD) stage, the instruction from the instruction cache is received (ø1) and then registers are read from the register file (ø2). During ø1 of the ALU stage, calculations are performed and branch results are checked while in ø2 a Load/Store instruction access to the data TLB (DTLB) is done. The Memory access stage during ø1 sends the physical address to the data cache while in ø2 the tags are compared and the data is returned from the cache. In the last stage, Write-Back (WB), results are written into the register file during its only phase, ø1.

The MIPS R3000 32-bit Processor contains 32 32-bit wide general purpose registers (GPR) and 32 co-processor zero (CP0) registers which enable the OS to access the TLB and page table, handle exceptions, etc. In addition, there are 3 special registers which are the program counter (PC), Hi, and Lo. The Hi and Lo registers store the double word (64 bits) result of an integer multiply or

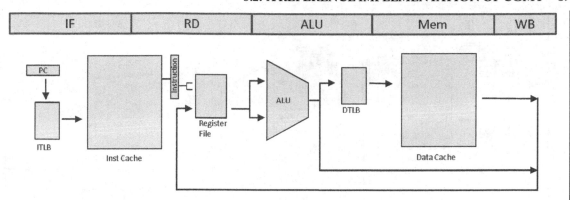

**Figure 3.1:** The MIPS R3000 Pipeline. This pipeline does not support multithreading.

divide operation. In the case of a divide, the quotient of the result is stored in the Lo register and the remainder in Hi. The GPR0 register is hard wired to the value '0'. The R3000 has single level instruction and data caches. Caches are write-through and direct mapped with physical indexing and physical tags. The caches can be accessed with a single cycle latency on a hit. Additionally, it features a memory management unit (MMU) with fully associative TLBs. The R3000 TLBs do not contain a PID, so they must flush on a software context switch.

We will expand this architecture (Figure 3.2) to include four (coarse-grain) hardware contexts and enable context swaps to be performed quickly in hardware. To achieve this we need to replicate all of the general purpose registers (GPR0-GPR31) for each hardware thread for a total of 128 registers. We will also need to replicate the Hi, Lo, PC, and most of the CP0 registers. In the case of the PC, we will assume there is a single *active PC* which is accessed each cycle; thus, on a context switch from context 1 to context 3, for example, the active PC will be stored to PC[1] and the value of PC[3] will be loaded into the active PC.

In addition to replicating existing register state, we will also add just a few new registers. We will have a 2-bit hardware thread ID (HTID). The HTID identifies which program counter is currently in operation, which partition of the register file is being accessed, and which entries in the TLBs are valid for this thread; that is, which hardware context currently owns the pipeline. In this model of multithreading, a single small register is sufficient to identify the thread because every instruction in the pipeline belongs to the same context.

To facilitate thread management at context switches, we will add two 4-bit vectors, the valid vector (VV) and the waiting vector (WV). The valid vector indicates which of the hardware contexts contain active threads (have current thread state stored in them). The waiting vector tells us which threads/contexts still have a memory operation in progress. Thus, a hardware context is only a viable target for a context switch if the corresponding bit in the valid vector is high and the same bit in the waiting vector is low.

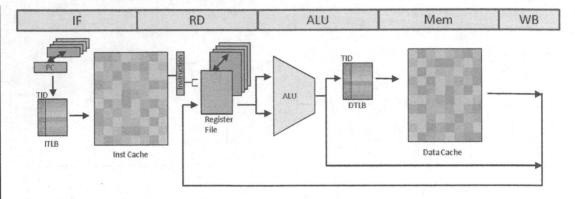

**Figure 3.2:** The MIPS R3000 Pipeline with support for Coarse-Grain Multithreading.

## 3.2.1    CHANGES TO IF

The fetch stage is responsible for accessing the instruction cache (ICache) at the address pointed to by the program counter. Our CGMT processor will have the primary *active PC*, and four *thread PCs*. On a context swap, the current active PC will be saved into the thread PC pointed to by the HTID, and one of the other thread PCs will be loaded into the active PC.

Because the ICache is accessed by physical address, it need not change; however, the ITLB must be able to translate the active PC into the correct physical address. Thus, each entry in the ITLB will contain a thread ID (TID) entry which identifies the hardware context to which it belongs. The ITLB will be discussed in more detail in Section 3.2.4.

When an instruction cache miss is detected (in phase ø1 of the RD stage), a context swap must be initiated. To do so, a signal will be sent to the IF stage by the RD stage to invalidate the current fetch stage and trigger a context swap. It will set the Waiting Vector bit for this context and update the active PC to point to another non-waiting context. It is unlikely that this will change the timing of the IF stage—for normal operation, the only new computation that takes place is the comparison of the HTID with the ITLB TID, which can easily be done in parallel with the page number comparison. The only other time we will initiate a context swap is for a data cache miss, as explained in the next section.

## 3.2.2    CHANGES TO RD

The main change in this stage will be simply to quadruple the size of our register file. Several organizations are possible, since we need only have a single register partition accessible at once. The simplest implementation would just have a single large register file, where the HTID bits are appended to the logical register name to identify the physical register. Alternatively, the register file could be physically partitioned in some way, with the HTID identifying the currently active partition. For simplicity, we will assume a large register file with 128 general purpose registers. The cost of increasing the register file by four times does not critically affect the complexity of the design, in this technology. It may affect register file latency, but we assume that in this technology it is still accessible in a cycle – SPARC processors of similar vintage had a large register file and similar pipeline. As we move to superscalar processors with high register port demands, and more aggressive processor technology, it becomes more likely that multiplying the size of the register file adds latency (e.g., an extra pipeline stage) to the register file access.

As previously mentioned, ICache misses are detected in ø1 of this stage, at which point a control signal should be triggered that will invalidate both the RD and IF stages and cause a context swap.

### 3.2.3 CHANGES TO ALU

No changes are needed to support coarse-grain multithreading with respect to the operation of the ALUs. The DTLB is also accessed in this stage, but is discussed in Section 3.2.4.

### 3.2.4 CHANGES TO MEM

The timing of both phases of this stage remain the same. In the case of a cache hit, this stage will behave the same as a non-multithreaded version. However, in the case of a cache miss, a memory request is initiated as before by putting the request into the load/store buffer, but we now need to have a separate load/store buffer per context. Thus, with four contexts we will have four load/store buffers, each one dedicated to a particular context. In this architecture, each load/store buffer is a single entry, since we context switch immediately on a cache miss, making it impossible to have multiple outstanding misses per thread. If we instead chose to switch on the use of cache miss data, then we could support per-thread hit under miss [Farkas and Jouppi, 1994, Kroft, 1981] or even miss under miss, and require multiple load/store buffer entries per context. Also, if we had multiple levels of cache hierarchy, we could have multiple L1 misses before an L2 miss causes a context switch, again necessitating more load/store buffer entries. Note that the cache, as described, must be non-blocking (support miss under miss) globally, even if each thread sees a blocking cache locally. A globally blocking cache on a multithreaded architecture would disable any performance gain due to multithreading.

When a data cache miss is detected and a context swap must be initiated, this stage will send a squash signal to the ALU, RD, and IF stages to invalidate the instructions in those stages, and also signal the IF unit to swap contexts (elaborated in section 3.2.6). Finally, we must set the corresponding bit for this hardware thread in the WV to indicate that it is now waiting on memory and hence not ready to continue executing. A request that returns from memory must update the register pointed to by the TID and destination register of the matching load/store buffer entry and clear the WV bit for that TID.

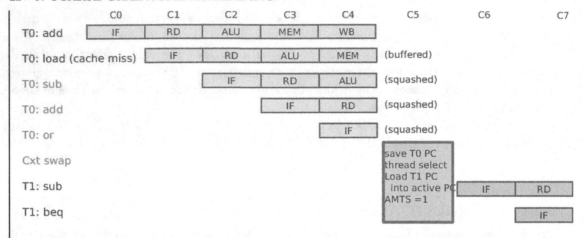

**Figure 3.3:** The operation of a CGMT context swap in our reference design.

**TLB Considerations**

Each entry in the ITLB and DTLB will contain a thread ID (TID) entry that identifies to which hardware context it belongs. Thus, we only get a hit in the TLB if the virtual page number (VPN) matches and the TID of the entry is the same as the HTID. We will only flush the TLBs (or the entries of the TLB corresponding to a particular TID) when we do a software context switch, swapping a thread out of a physical hardware context and replacing it with a new thread. We must have TIDs in the TLBs because we cannot afford to flush them on a context switch, since context switches are much too frequent in a CGMT architecture. Since four contexts now share each TLB, we may want to increase their sizes to maintain performance. We could either allow threads to share the TLBs using normal replacement policies, or partition the TLBs the way we do the register file, giving each context an equal portion.

## 3.2.5 CHANGES TO WB

Similar to the RD stage, the only change needed for the WB stage is the concatenation of the context ID with the logical register number to specify the physical register number in the register file.

## 3.2.6 SWAPPING CONTEXTS

Our CGMT processor will swap contexts every time there is a cache miss. To enable fairness, a reasonable thread selection strategy would be to always choose the next thread in sequence (round robin) which is both valid (VV[HTID] bit set) and not waiting upon memory (WV[HTID] not set). If all of the valid contexts are waiting on memory then no context switch is performed.

The sequence of pipeline events is shown in Figure 3.3 for the case of a data cache miss. In this case, the miss is detected in the MEM stage, and the context swap can be initiated in the following cycle. This will cause the squash of all instructions in the pipeline behind the load. This squash is safe because no instructions change program state (except the PC) before the MEM stage. The only possible older instruction would be in the WB stage, and would complete in the same cycle the load miss is detected; thus, we can change the HTID the following cycle without confusion.

Several things must happen to complete the context swap, but all can easily be accomplished in a single cycle. The active PC must be restored to point to the instruction immediately after the load, and that value is stored to the per-context PC corresponding to the current HTID. We will also set the WV[HTID] bit to indicate that this context is waiting for memory and not yet ready to execute. We will then select a new context to execute, if possible. The HTID is changed to point to the new context, and the PC at the index of the new HTID is then loaded into the active PC.

The cycle after this swap is performed, the IF stage will begin fetching from the newly swapped in context. Consequently, there are a total of four cycles lost due to a context swap caused by a DCache miss (IF, RD, and ALU stages are squashed and one extra cycle for the updating of the active PC, context PCs, and HTID). When the context swap is triggered by an instruction cache miss, we can limit the cost to a total of three cycles—the IF and RD stages are squashed and one extra cycle for updating the PCs and HTID.

While the pipeline and even the caches are only accessed by one thread at a time, the memory subsystem can have accesses from multiple threads active at once. Thus, memory requests must be identified by thread identifier (TID), for example in the miss status handling register (MSHR). Once the load request that generated the context swap comes back from memory, we will complete the original load instruction that triggered the swap. We cannot use the HTID, because that now corresponds to a different context; we will use the TID specific to that memory request, along with the original logical destination register, to identify the physical register. We can update the register either by waiting for a register port to be free, or by buffering the value in the load/store queue and doing the register write in the cycle after we swap back to this context, because no valid instruction will be in the WB stage using the register write port in that cycle.

### 3.2.7    SUPERSCALAR CONSIDERATIONS

This architecture extends simply to a superscalar implementation. Since the pipeline only executes instructions from one context at a time, we would make the same changes to a superscalar pipeline to arrive at a CGMT superscalar machine. The pipeline would drain the same way, context swaps would not change, etc.

## 3.3    COARSE-GRAIN MULTITHREADING FOR MODERN ARCHITECTURES

Coarse-grain multithreading has a clear place in the historical hierarchy of multithreading models and architectures. However, it is also a viable option in some modern designs.

CGMT's main advantage over single threaded designs, of course, is its ability to perform fast context swapping and hide memory latencies. But it shares that advantage with its cousins, fine-grain and simultaneous multithreading, which each have the potential to hide latencies more completely. The cost to achieve this gain is additional storage—a full hardware context per thread that we would like to be able to run. The hardware context includes a separate register file, PC, and any other needed contextual state information of the software thread. This cost is incurred in some way for each of the multithreading models.

Where CGMT has an advantage over the other models is that it deviates the least from existing single-thread processor pipelines. Thus, in a design where the starting point is a prior generation of the architecture without multithreading support, or perhaps an IP core without multithreading, relatively small changes are required to transform that architecture into one that can tolerate long latencies and keep the pipeline full. In an environment where the most critical latencies are long, and in between those latencies per-thread instruction-level parallelism is sufficient to keep the pipeline full, CGMT performance can approach that of the other multithreading models with lower implementation cost.

Coarse-grain multithreading does have some challenges. CGMT, like other multithreading models, typically improves system throughput. In the vast majority of cases, this improves the expected response time of each thread running in the system. However, there are some negative effects that can mitigate those gains, or in extreme cases cause performance degradation. In CGMT, threads are unencumbered in their use of the pipeline; therefore, the only negative interference is from competition for the caches, branch predictor entries, and TLB entries. Because those effects are shared with the other multithreading models, they will be discussed later in this book, in Chapter 6.

# CHAPTER 4

# Fine-Grain Multithreading

A Fine-Grain Multithreading architecture is one in which an instruction (or instructions, but most FGMT processors are scalar or VLIW) from a different thread can be issued every cycle, but only a single thread can issue an instruction in a particular cycle. That is, the architecture can perform a context switch in one cycle—in many cases, the pipeline actually seeks to do a context switch between threads *every* cycle. Therefore, on a pipelined machine, the FGMT processor would have instructions from multiple threads in the pipeline at once.

Like the CGMT processor core, this fine-grained core must also maintain the state of multiple programs in the machine at once; however, it is no longer acceptable for that state to be *nearby*, or accessible within a few cycles. In a fine-grain multithreaded machine, the state of all contexts (e.g., the program counter and registers) must be fully available to the pipeline. Because a FGMT processor can context switch in a single cycle, it can hide very short latencies (e.g., floating point operations of a few cycles and L2 cache hits) in addition to the long memory latencies that CGMT can hide.

## 4.1 HISTORICAL CONTEXT

In the 1960s, we were beginning to see a troubling imbalance between processor speeds (e.g., 100 ns for the execution unit) and memory access latency (1000 ns). This 10x time discrepancy led visionary Seymour Cray to introduce a barrel-shifting architecture in the peripheral processors of the CDC 6600 [Thornton, 1970]. That was the first implementation of FGMT. That architecture used ten independent peripheral processing units (PPUs) to handle memory accesses for the programs running in each PPU. However, all of the PPUs shared a single execution unit which ran at 10X the speed of the PPUs and was available for each PPU every tenth cycle. This allowed the execution unit to run at full speed, yet fully hide the latency of memory access.

The Denelcor HEP (Heterogeneous Element Processor) [Smith, 1982] was designed to solve complex numeric problems (namely nonlinear ordinary differential equations); in particular, they wanted to compete with the speed and parallelism of analog machines. They therefore used a combination of highly pipelined arithmetic units, aggressive multithreading, and aggressive clocking enabled by the elimination of bypass logic to achieve high computational throughput. The Tera architecture [Alverson et al., 1990], like the HEP architected by Burton Smith, also featured many threads and no bypass logic, but each instruction was a 3-wide VLIW.

Delco Electronics (a division of General Motors) created the first FGMT system targeted at real time constraints, called the Advanced Timing Processor, in 1984 [Nemirovsky and Sale, 1992a]. This processor was widely used in the engine control module in GM automobiles beginning in 1987, and for well over a decade after that. This machine used multithreading to do two things critical to real-time execution. First, it could divide timing resources unequally (one thread gets every other cycle, 32 other threads shared the remaining cycles round robin). Second, FGMT with round-robin scheduling provides deterministic execution even in the presence of other threads and non-unit latencies.

While most early machines took advantage of the presence of threads to eliminate the bypass logic, that is not necessarily inherent to the FGMT execution model. The Sun Niagara T1 [Kongetira et al., 2005], for example, retains the bypass logic of a normal pipeline, allowing it to fully utilize the pipeline with far fewer threads. For instance, while the HEP had up to 128 hardware contexts per pipeline, the T1 has four.

Some of the earliest academic literature on fine-grain multithreading came from Ed David-son's group in the late 1970s [Kaminsky and Davidson, 1979]. They advocate a fine-grain multi-threaded, single-chip, pipelined microprocessor to maximize pin utilization and minimize pin and I/O overhead (compared to a chip multiprocessor).

## 4.2   A REFERENCE IMPLEMENTATION OF FGMT

We will build our FGMT processor on the same MIPS R3000 baseline architecture introduced in Chapter 3. Our first critical design decision is whether to retain the bypass logic. In keeping with the earliest FGMT designs, we will eliminate it and simplify the pipeline design. This requires that we support at least five threads to get full utilization out of the 5-stage pipeline—only four stages, RD-WB, interact via various forms of bypass, but if we neglect IF, then we risk introducing new instructions that could interact with instructions that may still stall. Even five threads may be insufficient, due to long-latency instructions, particularly cache misses. These stalls will cause the associated thread to become unavailable for more than five cycles. Thus, to increase the likelihood of finding a ready instruction to fetch (introduce into the pipeline) each cycle, we will support 3 additional threads for a total of 8.

With this architecture, no two dependent instructions can ever be co-resident in the pipeline. Thus, we do not need the bypass logic, as mentioned, or any other inter-instruction register depen-dence checking. However, same-address-space threads can still have memory-based dependences which are handled in a straightforward manner by maintaining the in-order pipeline.

Unlike the CGMT processor, we no longer have a single thread ID (HTID) which governs the entire pipeline; instead, each instruction carries its own TID (which is now 3 bits long to support the 8 total contexts) through the pipeline, thus adding three bits to various structures that hold in-flight instruction data. We still need the Valid Vector and Waiting Vector registers, now resized to each be eight bits long, where the VV is unchanged and the meaning of the WV is changed slightly. A thread's WV bit will be set once an instruction is fetched from a ready context and then reset once it completes. This ensures that the pipeline can never contain more than one instruction from a single context. The first thing to happen in the IF stage is to select the PC from the thread

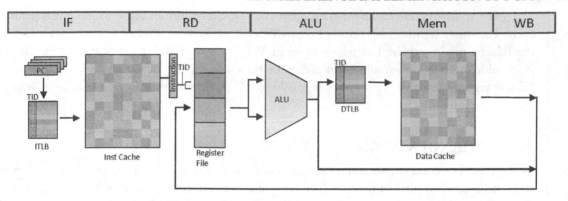

**Figure 4.1:** The MIPS R3000 Pipeline with support for Fine-Grain Multithreading. Although our target design would support 8 threads to maximize throughput, we show a 4-thread implementation for simplicity.

to be fetched next. It will choose a thread for which the associated VV bit is set and the WV bit is unset. If no such thread exists, it will not fetch an instruction in that cycle.

We will detail the required changes to the pipeline more fully in the following sections. The modified R3000 pipeline appears in Figure 4.1.

## 4.2.1 CHANGES TO IF

We introduce thread select logic and eight program counter fields to the fetch unit. While the thread select logic is not unlike that of the CGMT, in that architecture it was invoked rarely, while in FGMT thread select will be invoked every cycle. Thus, each cycle the unit will attempt to identify a thread which is active, but does not have an instruction in-flight, again using the two bit vectors, VV and WV. We will fetch from the corresponding PC, append that TID to the fetched instruction, and set the corresponding WV bit. The thread select function could enforce fairness by employing a simple round-robin priority mechanism, whereby the next fetched instruction is from the next context in sequence (after the last fetched context) which is active and does not have an instruction in flight.

Despite these additions, the fetch unit is potentially greatly simplified, because we can completely eliminate the branch prediction logic. In this model, branches are always resolved before the following instruction is fetched. Although the R3000 did not have branch prediction (due to the branch delay slot), that solution does not scale to deeper pipelines, while a fine-grain multithreading architecture can always forgo branch prediction if we prevent control-dependent instructions from entering the pipeline (as we do in this implementation).

As in the CGMT model, we do not change the instruction cache at all, but the ITLB will again need to support entries from multiple threads (identified by TID).

On an instruction cache miss, no instruction would be introduced into the pipeline for that cycle. The missing thread will already have its WV bit set, so it is sufficient to leave that bit set until the ICache miss is resolved, at which point its WV bit is reset and it can attempt to fetch again. A possible optimization would be to pre-probe the cache for the PCs of ready threads, when the ICache tags are available. This makes it possible to (1) not lose a cycle on each ICache miss and (2) start the ICache miss handling earlier.

## 4.2.2    CHANGES TO RD

In the CGMT processor, we could have exploited the fact that only a portion of the total register space need be available at a time, and we have time between threads to switch the contents of the accessible register file. In the FGMT architecture, registers for different contexts are accessed in quick succession, making such a design much more difficult. Thus, we (again) simply increase the total size of our register file. Our integer register file becomes 8*32=256 entries (in reality we only need 31 register per thread since R0 is hardwired to the valued 0) , and we also have eight versions of the CP0 registers and Hi and Lo. Physical registers are accessed by simply appending the TID to the high bits of the architectural register name.

With 256 registers, we may begin to be concerned about the access time of the register file, even though the port requirements are still minimal. Doubling the access time to a full cycle adds minimal complexity to this design, however. Much of the motivation for keeping access to a half cycle in the original R3000 design is to eliminate the need for bypass between the ID and WB stages. By writing in the first phase and reading in the second phase, a write and a read to the same register in the same cycle will be automatically resolved in the original pipeline. In our new pipeline, this hazard cannot occur. As a result, having both the register read in RD and the register write in WB take a full cycle does not create a data hazard or introduce bypass or conflict resolution logic. However, taking longer to access the register file would shift other operations in our pipeline. Thus, for the purposes of the following discussion, we assume that the register access time is unchanged.

## 4.2.3    CHANGES TO ALU

As in the CGMT implementation, no change to the ALU stage is necessary. The DTLB will support translations for multiple contexts, and will again be very similar to the DTLB for the CGMT design.

## 4.2.4    CHANGES TO MEM

The physically accessed cache in this design need not be changed in any way to support multiple threads, as the translations made in the DTLB will resolve differences between contexts.

To properly handle a cache miss, it is sufficient that a missing instruction does not complete immediately, keeping its context's WV bit set. This prevents a new instruction from being fetched. Because only one instruction from a thread can be in the pipeline at once, stalling an instruction cannot affect the memory ordering within a thread. Possible dependences between memory access instructions in different threads are handled naturally using standard mechanisms because instructions move through this pipeline in order.

As in the CGMT architecture, the cache must be non-blocking globally; however, there is no need to be non-blocking within a thread (e.g., to support hit under miss).

### 4.2.5 CHANGES TO WB

At the writeback stage, we will again append the per-instruction TID to the destination register to determine the physical register being written. In this stage, we will reset the bit in the WV bit vector corresponding to that instruction's TID, enabling that context to again be considered for fetch.

It should be noted that in order to take full advantage of the FGMT architecture's ability to hide memory latency, we must allow instructions to complete (enter the WB stage) out of order (more specifically, in a different order than they are fetched). This does not allow the reordering of instructions within a thread, since only one instruction can be in the pipeline. But it does allow non-stalled threads to bypass stalled threads, allowing those threads to complete instructions and continue to fetch while another instruction is stalled waiting for memory. So this introduces no ordering or dependence hazards (except those already being tracked by the load/store queue or MSHRs), but does introduce a possible resource hazard on the write port for the register file. This is because if we are doing a good job of keeping the pipeline full, there is a high probability there is an instruction in the WB stage in the same cycle that a missing load instruction returns its data from memory.

We can resolve this hazard by either adding a write port to the register file or by applying port-stealing—waiting until the write port is free. The write port will be free frequently, even for a well-utilized pipeline. Branch instructions and store instructions, among others, do not use the write port. Bubbles caused by dependences, I cache misses, D cache misses, or insufficient thread level parallelism will also free the port.

### 4.2.6 SUPERSCALAR CONSIDERATIONS

This architecture does not extend easily to a superscalar machine. The restriction to only introduce instructions from a single thread per cycle means that we now would have multiple instructions from the same thread in the pipeline. Thus, we now need (perhaps limited) dependence checking, as well as branch prediction. Presumably, we can still choose to eliminate the bypass logic, as we would not introduce instructions detected as dependent in the same cycle; however, we would need the ability to buffer some of the fetched instructions for later introduction after the first clears the pipeline, in the case of a detected dependence. Similarly, we can still eliminate branch prediction as long as we can detect or predict the end of a fetch block (i.e., the presence of a branch instruction).

This explains why multi-issue FGMT machines have typically been VLIW. Our architecture naturally extends to a VLIW architecture, where one VLIW instruction per context is allowed in the pipeline at a time. Most of the previous discussion in this Chapter is easily applied to that architecture, just by substituting "VLIW instruction" for "instruction."

## 4.3 FINE-GRAIN MULTITHREADING FOR MODERN ARCHITECTURES

Like coarse-grain multithreading, fine-grain multithreading has a clear place in the historical hierarchy of multithreading models. Perhaps even more than CGMT, it also has distinct advantages that still make it a viable option for modern processor designs.

In less aggressive embedded designs, for example, a scalar pipeline may be sufficient, making FGMT the most aggressive form of multithreading available, and possibly a good choice if non-unit latencies are frequent.

In the most energy-conservative designs, fine-grain multithreading enables design points not possible with other multithreading models or without multithreading. We can build a fine-grain pipelined processor without power-hungry pipeline logic (out-of-order scheduler, register renaming, dependence checking, bypass logic, etc.), but at the same time hope to keep the pipeline full to near capacity. This maximizes energy efficiency by both eliminating unnecessary structures and not wasting power on unused pipeline cycles.

We might make the same choice if we were heavily constrained by area, again benefiting from the ability to save logic yet still keep the pipeline full.

This choice also may be attractive from a pure performance perspective if the reduced logic allows us to crank up the clock rate sufficiently. Whether such an architecture could compete with an SMT processor would depend on the availability of threads and the total available instruction-level parallelism.

One disadvantage of this design choice, especially if we eliminate the bypass logic and restrict ourselves to one instruction per thread, is that even more contexts are required to keep the pipeline full. This means more PCs and more complex thread selection logic. Perhaps most important, it means that FGMT requires a large register file. The register file may become a critical path, or we may choose to pipeline register access, increasing the number of stages in the pipeline (and again increasing the number of contexts needed).

The second disadvantage of that design choice (eliminating bypass) is that single-thread performance suffers greatly. In this architecture (assuming the 5-stage pipeline) the peak throughput of a single thread is 1/5 that of the original single-thread pipeline. In the presence of sufficient threads, however, we are much more likely to achieve the full throughput of the processor (IPC of 1) with the multithreaded version than with the original pipeline.

On the other hand, if we combine FGMT with full support for bypass logic and inter-instruction dependencies, we need fewer threads to keep the pipeline full, resulting in smaller register files, etc. More importantly, we observe little or no slowdown for single-thread execution, unless the pipeline still needs to be expanded for increased register file access latency. Thus, this architecture with bypass logic sacrifices little performance in the worst case and exceeds non-multithreaded performance with only a few threads.

FGMT also shares many of the same challenges as other multithreading models: competition for cache space, TLB space, thread co-scheduling issues, etc. However, we do not yet see the per-cycle resource conflicts we will see with SMT. As mentioned previously, since so many of these issues are common, they will be addressed in Chapter 6.

CHAPTER 5

# Simultaneous Multithreading

Simultaneous Multithreading enables multiple threads to issue instructions to the execution units in the same cycle. In this way, instructions from other threads can be used to fill out the issue width of the processor when one thread lacks the instruction level parallelism to fully utilize the execution bandwidth.

Thus, it is impossible to identify "who" owns the pipeline, either across a window of cycles, or even for a single stage in a single cycle. All resources, then, must be accessible to the pipeline at all times—it is not necessarily advantageous to physically partition the register file, renaming logic, or instruction queues.

## 5.1 HISTORICAL CONTEXT

The 1990s saw the introduction of superscalar processors, while accelerating processor clock speeds drove memory latencies dramatically higher. This combination brought the gap between available hardware ILP and sustainable software ILP to a new high. In particular, the advent of superscalar execution cores introduced a new form of lost execution bandwidth—horizontal waste, as described in Chapter 2. For the first time, even single-cycle instructions could result in lost execution throughput. This created the need for execution models that could more aggressively supply instruction bandwidth, and in particular could hide even horizontal waste. Out-of-order execution would be introduced into many designs to help bridge that gap, but it could not close it completely.

Simultaneous multithreading takes its name from the ISCA paper in 1995 [Tullsen et al., 1995]. The simultaneous multithreading proposals in the literature that would most closely resemble the implementations that would follow appeared in the early to mid 90s [Gulati and Bagherzadeh, 1996, Hirata et al., 1992, Keckler and Dally, 1992, Serrano et al., 1993, Tullsen et al., 1995, 1996, Yamamoto and Nemirovsky, 1995, Yamamoto et al., 1994]. The first commercial implementations were announced in 1999 [Emer, 1999] and actually delivered in 2002 [Koufaty and Marr, 2003].

The first to announce a simultaneous multithreading design was the Digital Alpha 21464 (internal name EV8) [Diefendorff, 1999, Emer, 1999]. The 21464 was perhaps the most aggressive superscalar machine ever designed, as it sought to follow the recent rapid progression (of scalar to 2-wide superscalar to 4-wide superscalar) with an 8-wide superscalar, out-of-order pipeline. Such a design, if single-threaded, would be highly underutilized for most applications, with significant instruction bandwidth lost to horizontal waste. That design would feature support for four hardware contexts, and a pipeline implementation that was quite consistent with the earlier work co-authored by the University of Washington researchers and key members of the Alpha design

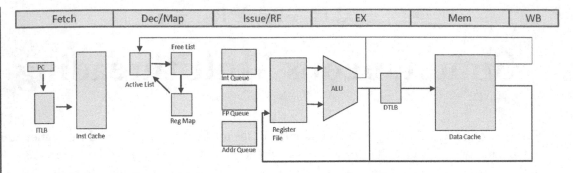

**Figure 5.1:** The pipeline of the MIPS R10000 processor, as seen by a load instruction.

team [Tullsen et al., 1996]. However, the Alpha architecture (along with IP and personnel) would become a victim of corporations and units changing hands. Digital was acquired by Compaq, who chose not to go into microprocessor production, after which the Alpha group was acquired by Intel, who was not seeking an alternative to its two existing processor and instruction set architecture families.

However, that did not prevent Intel from being the first to deliver an SMT processor. Intel's Pentium 4 Xeon server processors [Hinton et al., 2001, Koufaty and Marr, 2003] introduced simultaneous multithreading under the proprietary name Hyper-Threading or Hyper-Threading Technology. Because the Pentium 4 is closer to a 3-wide superscalar machine, it did not require as aggressive an implementation of SMT. It featured two hardware contexts, and chose relatively straightforward implementations of the instruction queues—rather than being fully shared, they were either statically partitioned into two equal halves (one for each thread in multithreaded mode) or combined for a single partition (in single threaded mode).

When Intel introduced multicores, they stepped back to less complex core designs to save real estate and enable multiple cores on a die, and for a period multicore and multithreading were mutually exclusive options. As a result, IBM was the first to introduce an SMT multicore product with the Power5 [Kalla et al., 2004]. Eventually, Intel was able to find the real estate for both in the Nehalem processors [Dixon et al., 2010].

## 5.2   A REFERENCE IMPLEMENTATION OF SMT

For the purposes of this chapter, we prefer to add our SMT design to a processor that supports superscalar, out-of-order execution. Thus, instead of the R3000 baseline used in the previous two chapters, we will use the MIPS R10000 [Yeager, 1996] pipeline as our baseline here. Figure 5.1 shows the baseline R10000 pipeline. The MIPS R10000 is a four-wide, dynamically scheduled superscalar processor. It also has physically tagged (although virtually indexed) caches. Out-of-order scheduling is enabled via the combination of *register renaming*, the *active list*, and three *instruction queues* (integer, floating point, and address queue for loads and stores).

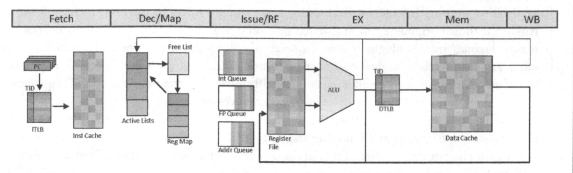

**Figure 5.2:** The MIPS R10000 pipeline with support added for simultaneous multithreading.

Register renaming is implemented with a *register map* (of architectural to physical register mappings) and a *free list* (physical registers not currently assigned to an architectural register). Not only does register renaming enable the elimination of false register dependences, but it is also a critical component to enable out-of-order issue. It allows the pipeline to separate register write from commit, which is critical for out-of-order execution—instructions will complete out-of-order and before it is even known if they will commit. Thus, an instruction writes to the physical register file as soon as it completes execution, but does not commit until it is graduated from the active list.

The active list maintains the ordering of instructions and holds the name of the physical register that will become obsolete if the particular instruction commits (the previous mapping of the instruction's destination register). This ordering is useful for branch mispredict and exception recovery, and of course enables in-order commit. When an instruction commits, it is graduated from the active list and the physical register held in its active list entry is placed on the free list.

The instruction queues hold instructions waiting for their operands to become available. Once all their operands are available, instructions are marked as ready and the scheduler will select ready instructions for issue to the execution units. An instruction is marked ready when its operands will either be available in the register file or the bypass network by the time the instruction gets to the appropriate (e.g., EX for arithmetic instructions) pipeline stage.

In this pipeline diagram, *commit* is not explicitly shown as a pipeline stage, but happens some time (possibly many cycles if a prior instruction is stalled) after WB. It utilizes the active list and the free list, which are shown here in the Dec/Map stage. An instruction can commit when it has completed with no exceptions or misspeculation, and all prior instructions have committed.

The pipeline modified to support SMT is shown in Figure 5.2. There are two reasons why most research and commercial implementations of SMT pair it with out-of-order execution. The first is simply timing—most high-performance processors were either out-of-order or transitioning to out-of-order when SMT was being seriously considered by industry. The second is more important, and perhaps influenced the first – simultaneous multithreading is highly synergistic with out-of-order instruction issue. The same hardware that selects instructions for issue based on operand availability, ignoring instruction fetch order, also naturally mixes instructions each cycle with no regard for thread origin—again, it is based only on operand availability. As long as operand (i.e., register) names are unambiguous across contexts, the dependence tracking and scheduling are the same. As a result, the marginal cost of implementing a simultaneous multithreaded processor on top of an out-of-order processor core is significantly lower than the marginal cost of implementing simultaneous multithreading on top of an in-order processor.

Thus, this figure also assumes that the full mixing of instructions (the S in SMT) happens in the instruction queues and following stages; there is limited per-cycle sharing (or possibly no sharing) of pipeline stages prior to that in the pipeline. For example, fetching from a large number of threads in a single cycle would require excessive new complexity. Thus, although the fetch unit has the full set of possible program counters to choose from each cycle, we assume that it will select a subset (e.g., one or two) to fetch from each cycle. It should be recognized, however, that although increasing the number of threads fetched per cycle increases the complexity of the instruction cache, it can be used to decrease the complexity of the rename logic. That is, the number of dependences the rename logic must account for is quadratic in the maximum number of instructions fetched *per thread*. Other advantages of fetching multiple threads per cycle are discussed in Section 6.3.

The SMT architecture, then, again must have a larger physical register file. It will have four (for this implementation) program counters. Although we must fill much more instruction bandwidth than the FGMT machine from the previous chapter, because our base architecture is superscalar, the ability to utilize multiple instructions from each thread as well as multiple threads per cycle means that we can actually get away with fewer contexts. Again, there is no central thread ID, but a TID carried by each instruction. However, that TID is primarily used in register renaming. For most of the pipeline between Map and Commit, the logic depends on the physical register name and is largely thread-agnostic. We will assume we fetch from one context per cycle, to minimize changes to the ICache and branch predictor.

More specific changes to the pipeline are described below.

## 5.2.1   CHANGES TO FETCH

Again, physically tagged caches imply no significant changes to the instruction or data cache. The branch predictor is unchanged, although like other multithreading models, SMT places more pressure on the branch predictor, possibly resulting in lower accuracy. This is discussed further in Section 6.2. As in CGMT and FGMT, the TLBs must support mapping entries for multiple threads. The thread select logic may look similar to that of the FGMT reference architecture since we only fetch from one thread per cycle; however, in the context of an out-of-order machine, there is significant opportunity to improve performance by carefully selecting the thread to fetch each cycle. This is because some threads are more likely to bring in instructions that would hold critical resources for a longer period, or that are more likely to get squashed. This is discussed extensively in Section 6.3.

The thread ID (TID) of the selected context will be appended to each fetched instruction as it proceeds into the pipeline.

Other architectures, such as the Power5 [Kalla et al., 2004], take an alternative approach to thread selection and fetch. Thread select takes place after fetch in the pipe, choosing threads to send to decode from among the per-thread instruction buffers. In that model, the only job of fetch is to identify the least full buffer each cycle and fetch into that buffer.

If we instead chose to fetch from more (e.g., two) threads per cycle in our implementation, to reduce the fetch fragmentation caused by branches [Tullsen et al., 1996] and thus increase fetch throughput, our thread select logic would need to select two threads for fetch each cycle and our instruction cache would need to be dual ported.

## 5.2.2  CHANGES TO DEC/MAP

Register renaming, among its other advantages, allows us to naturally disambiguate registers from distinct threads, completely eliminating that concern. That is, we need do nothing special to tell R4 from thread 0 from R4 from thread 2, because they will have different physical names after renaming. Thus, as instructions move further through the pipeline, neither the scheduler nor the register file need reference the TID. They operate purely based on physical register names, as in the original out-of-order pipeline. The TID will be used in the Dec/Map stage to do register renaming, then for the most part be ignored except for access to the DTLB, and then finally to identify the correct active list partition when the instruction is ready to commit.

The register renamer must still record mappings for all architectural registers, and our architectural registers have expanded by a factor of four. Thus, we append the TID to the architectural register name when looking up the physical mapping in the now larger register map. This makes register renaming a potential critical timing path, but is likely still a good trade-off—this architecture scales the register map tables (which are small) by a factor of four, but does not scale the register file (much larger) by the same factor.

The active list also requires per-context partitioning. We assume a static partitioning. We could also consider a dynamic partitioning of the active list, which would allow a thread to temporarily use more space when needed, or allow the architecture to allocate more space to a thread that showed evidence that it would make better use of the space than others. However, it is critical in a multithreaded machine that one stalled thread cannot completely take over a resource that would cause other threads to stall. The easiest way to ensure this is to use equal static partitions.

## 5.2.3  CHANGES TO ISSUE/RF

The issue queue is largely unchanged, because the wakeup logic (the logic responsible for waking up instructions waiting for their source operands to become available) depends only on the physical register names. The select logic only needs to know which instructions are ready. Physical register names are a couple bits wider because the physical register file is larger. The only new logic is the ability to do per-context flush on a branch mispredict or exception. However, the queues already have the ability to do selective flush. The baseline must identify instructions with an instruction ID higher than (i.e., fetched later than) the mispredicted branch. Thus, to identify instructions from context C with an instruction ID higher than the branch is only slightly more complex, especially since the R10000 stores instructions in the queues completely unordered (that is, not in age order).

The register file must grow to account for the increase in architectural registers. But it will not grow by a factor of four. For example, the integer register file in the baseline design has 64 registers. 33 of those physical registers will always contain committed architectural register values (GPR1-31, Hi, Lo), allowing the machine to support 31 instructions in-flight (renamed but not committed) that have an integer destination register that has been mapped. To support four hardware contexts and the same number of in-flight instructions would require $4 * 33 + 31 = 163$ registers, an expansion over the original register file of just over 2.5x.

A larger register file would certainly take longer to access, although in this figure we assume it can still be done without changes to the pipeline. Because the R10000 baseline actually reads and writes registers in a half clock period, it is reasonable to assume that the larger register file could be accessed in a full clock cycle. Even if the increased register file latency requires a full extra pipe stage, Tullsen et al. [1996] show that an extra register stage costs single-thread execution 2% on average. The Alpha 21464 pipeline demonstrated no extra pipeline stages resulting from the larger register file.

## 5.2.4  OTHER PIPELINE STAGES

Similar to CGMT and FGMT, the data cache is unchanged, and the DTLB is supplemented with TIDs. The commit logic, which manages the active list, freeing active list entries and registers, should still enable the commit of four instructions per cycle. It is a design decision whether we only consider one context per cycle for possible commit, or enable instructions from multiple active list partitions to commit in a single cycle. The latter makes commit less likely to become a bottleneck because it will maximize commit throughput. Because commit is a lightweight mechanism in this architecture (it simply moves register names to the free list), the extra complexity of selecting instructions for commit across multiple partitions is unlikely to affect the clock cycle.

## 5.3  SUPERSCALAR VS. VLIW; IN-ORDER VS. OUT-OF-ORDER

Much of the research, and certainly the most visible commercial implementations of simultaneous multithreading, are built around a superscalar, out-of-order execution core. This is because the out-of-order issue logic greatly minimizes the marginal logic necessary to enable SMT. Similarly, the superscalar's ability to dynamically find and exploit parallelism each cycle (whether it is found in-order or out-of-order per thread) is a much better match for the goals of SMT than VLIW, where the parallelism is constrained by compiler decisions. However, several proposals have combined either in-order superscalar or VLIW execution with SMT, including much of the earliest SMT research.

In the mid-1990s, when the first SMT research appeared in the literature, the industry had not yet made the transition to out-of-order microprocessors, although significant research indicated the viability of that direction. As a result, most of the initial papers assume in-order cores as a baseline, and evaluate the performance gain of simultaneous multithreading over the in-order single-thread core. This includes Gulati and Bagherzadeh [1996], Tullsen et al. [1995], Yamamoto and Nemirovsky [1995]. Simultaneous multithreading is actually *more* effective (from a performance perspective) when combined with an in-order core. This, again, is because the gap between hardware parallelism and software utilization of that parallelism is actually maximized on an in-order superscalar. Out-of-order execution closes that gap somewhat, reducing the maximum utility of SMT. It's been shown, for example, that in a highly threaded environment (e.g., 4 active threads), the marginal gain of out-of-order execution vs. in-order is very low compared to the same comparison on a single-threaded machine [Hily and Seznec, 1999]. The Clearwater Networks XStream architecture implemented SMT on an 8-wide superscalar in-order processor.

Although there are no commercial implementations of SMT paired with VLIW (very long instruction word) processing, several research works describe such an implementation. The combination of SMT and VLIW can take two forms, which we will call *internal SMT* and *external SMT*. In internal SMT VLIW, individual operations of a single per-thread VLIW instruction are cocheduled with other threads' operations onto a set of functional units (where the makeup of the hardware units now need not match the makeup of the VLIW instruction). This allows the processor to maintain the simple semantics of VLIW execution, yet still tolerate latencies to individual operations without stalling the pipeline, as in an out-of-order machine. Examples of internal SMT architectures include Keckler and Dally [1992] and Hirata et al. [1992].

Conversely, external SMT VLIW still sends each VLIW instruction through the pipeline as a single atomic unit. In a processor that can issue multiple VLIW instructions per cycle (such as the Itanium architecture [Sharangpani and Arora, 2000]), the SMT version would have the ability to issue VLIW instructions from multiple threads per cycle. Thus, this architecture is very similar to an in-order superscalar SMT, except that each instruction being considered for issue is a VLIW instruction. This architecture is explored in Wang et al. [2002].

# 5.4   SIMULTANEOUS MULTITHREADING FOR MODERN ARCHITECTURES

Based on the number of high-profile implementations of simultaneous multithreading processor cores, it is clear that SMT is a relevant architecture for modern designs. But it is still worth understanding where in the design space it makes most sense.

In the most performance-intensive environments, no architectural alternative pulls as much performance (particularly when defining performance as instruction throughput per cycle) out of a particular pipeline as SMT.

However, in energy-constrained applications, there is also a place for SMT. If we combine tight energy constraints with abundant thread level parallelism and high performance demands, we might combine SMT with wide in-order cores. This architecture would allow us to get high throughput with multithreading without the power cost of the dynamic scheduling hardware. We might even reduce the area and power devoted to branch prediction.

If we are less confident about the abundance of threads, but still aware of energy and performance, we might add the out-of-order logic back in. This architecture uses more power, but the resulting high utilization of the pipeline amortizes that power over a higher rate of instructions completed per cycle.

When compared to a chip multiprocessor, in particular, an SMT processor can be highly energy efficient. Assuming the same per-core instruction width, the cost of running one thread on an SMT core is only marginally higher than on a non-SMT core of the CMP; however, the cost of enabling the second thread on the SMT core is much lower than enabling the second thread on a CMP [Seng et al., 2000].

A simultaneous multithreaded processor shares execution resources at an even finer granularity than either coarse-grain or fine-grain multithreading. Thus, it inherits all of the resource management issues of the other multithreaded architectures (shared caches, shared fetch bandwidth, shared branch predictor, etc.) and creates even more (contention for execution units, per-cycle cache bandwidth, instruction queue space, renaming registers, etc.).

Again, since many of these contention and management issues are common to all models of multithreading, they will be discussed in the next chapter.

# CHAPTER 6

# Managing Contention

Significant research, both in academia and industry, investigates how to properly design, architect, and generate code for multithreaded processors. The most profound difference between multi-threaded architectures and other processor architectures (e.g., time-shared uniprocessors or multi-processors) is the level of sharing of processor resources. This sharing can be a two-edged sword. On one hand, it increases utilization, energy efficiency, and area efficiency. On the other, it allows co-scheduled threads to conflict over a wide set of resources.

Although threads conflict over shared resources in many processors, including chip multiprocessors, the fine-grain resource sharing in multithreaded processors greatly increase the ability for one thread to interfere with another. With an SMT processor, for example, threads compete with other threads *every cycle* for instruction and data cache space, instruction and data cache bandwidth, fetch and decode bandwidth, instruction queues, renaming registers, issue/execute bandwidth, re-order buffer entries, etc. As a result, there is significant research both in academia and industry centered around managing this contention. If we can limit the negative performance impact of that contention, we can even further increase the utility of multithreading. This chapter focuses on managing potentially negative interference. The next chapter examines some opportunities for exploiting positive interference between threads.

The goal of multithreading is always to maximize throughput by preventing a stalled instruction, or even a completely stalled thread, from stalling the processor. Thus, the goal of resource management is typically to minimize the resources allocated to, or occupied by, a stalled or soon-to-be-stalled thread. This then maximizes the throughput of the unstalled threads. Other times, we seek to identify which threads would make the most effective use of a contended resource.

Specifically, in this Chapter we will examine cache contention (Section 6.1), branch predictor contention (Section 6.2), fetch policies to manage contention (Section 6.3), register file contention (Section 6.4), operating system scheduling around contention (Section 6.5), compiling around contention (Section 6.6), contention caused by synchronization (Section 6.7), and contention-related security issues (Section 6.8).

## 6.1 MANAGING CACHE AND MEMORY CONTENTION

Multithreading simultaneously increases the rate of cache misses (due to increased contention for cache space) and increases the processor's ability to tolerate misses. In most cases, the latter effect dominates; however, even when memory latency is being effectively tolerated, careful design of the memory hierarchy is still warranted. First, because the optimal configuration for a multithreaded system is typically quite different than a memory hierarchy optimized for single-thread execution. Second, the more effectively we solve the memory latency problem, the more pressure we put on the throughput of the memory hierarchy.

While it would be possible to statically partition the private caches of a multithreaded processor and eliminate inter-thread contention, there are several reasons why we prefer a single cache shared among all contexts [Nemirovsky and Yamamoto, 1998, Tullsen et al., 1995]. First, we expect dynamic sharing of the cache to usually be more efficient than static partitioning because threads naturally adapt their use of the caches to the size of their working sets. Second, cooperating threads from the same address space can share data in the single cache. Third, and perhaps most importantly, a shared cache maximizes the utility of the cache at all levels of multithreading—that is, when many threads run they dynamically share the cache, and when a single thread runs it naturally expands to fully utilize the single cache. The Intel Pentium 4 architects compared the performance of a partitioned cache vs. a fully shared cache on the Pentium 4 hardware [Koufaty and Marr, 2003]. They found the shared cache outperformed the partitioned cache, averaging about a 12% gain.

However, there will always be interference in the cache between threads. Sometimes it can be positive, if the threads share code or data. Often it is negative interference, as one thread causes lines of another thread to be evicted. Thus, there are several tangible advantages to partitioned caches. They provide isolation between threads and improve performance predictability. With an unmanaged shared cache, performance of one thread can vary dramatically depending on the nature of the threads with which it is co-scheduled, and how they use the shared cache space. It is natural, then, to seek the best of both worlds: the global efficiency and adaptability of a shared cache, but the predictability and performance isolation of partitioned caches.

Cache contention can be addressed in two ways: (1) by managing which threads are co-scheduled onto a shared cache, and (2) by managing the cache being shared. We will discuss scheduling issues separately in Section 6.5 and focus on the latter architectures here. Some multithreaded cache research seeks to maximize global performance (letting one thread get slammed by another thread rarely provides the best global solution), while other research views the providing of performance isolation as a goal in itself.

Two of the earliest works on caches for simultaneous multithreaded architectures made similar fundamental observations [Hily and Seznec, 1998, Nemirovsky and Yamamoto, 1998]. Both showed that multithreaded caches tend to be more sensitive to both associativity (not surprisingly, preferring more associativity) and linesize (preferring smaller line sizes) than single-threaded cores. Both parameters allow the caches to better address the added pressure of multiple instruction streams. In addition, Hily and Seznec [1998] show that a multithreaded processor can put far more pressure on the bandwidth of the L2 cache than a single-threaded processor. This comes from a combination of the reduced locality and the ability to tolerate those misses and continue to generate memory accesses at a high rate.

López et al. [2007] examine cache designs that can dynamically trade off capacity for latency. Among other things, this allows single-thread execution and multithreaded execution to find different design points. Not surprisingly, the optimal point for the two can be quite different—with more pressure on the cache capacity and more latency tolerance, the multithreaded core prefers larger, slower caches.

With some hardware-supported control over the regions, ways, or lines of the cache that each thread can occupy [Settle et al., 2006, Suh et al., 2002], we can achieve the best of both worlds (partitioned and shared caches). We can place tight constraints on threads (effectively creating private caches) when isolation is a priority or when one or more of the threads do not share well, and loosen those constraints when sharing space dynamically is preferred. In fact, we can do even better by allocating the private regions unevenly, giving more cache to applications whose performance is more sensitive to cache size.

Zhu and Zhang [2005] show that many of these same effects remain true as we move further from the core and out to DRAM, as the best memory controller policies favor throughput over latency. This is again because SMT creates more misses yet tolerates long latencies. They also show that scheduling policies that are not thread-oblivious, but consider the microarchitectural state of competing threads, can improve performance.

In summary, then, we see that the best cache design for a multithreaded core would be larger and more associative, possibly with smaller cache lines, than a single-threaded cache. In fact, we would gladly trade those features for slightly increased latency. A multithreaded cache will both incur more misses/access (due to conflict between threads) and potentially significantly more accesses/second (due to increased latency tolerance). This greatly increases misses/second sent on to further levels of the cache hierarchy, requiring careful attention to the bandwidth and throughput of the cache/memory hierarchy and interconnect.

## 6.2    BRANCH PREDICTOR CONTENTION

Very similar to caches, a multithreaded processor can simultaneous incur more branch mispredicts (due to inter-thread contention), yet also be more tolerant of the number of branch mispredicts.

A fine-grained multithreaded processor that does not allow multiple instructions from the same thread into the pipeline at once can eliminate branch prediction altogether. For all other architectures, branch prediction logic in the fetch unit could either be partitioned per context or simply aggregated. The advantage of the latter is that the full size of the branch predictor is available when only a single thread is executing, making this the design of choice with few, if any, exceptions. Running multiple threads simultaneously puts additional pressure on the branch prediction logic—this is because the "working set" of important branches is multiplied by the number of threads.

However, SMT also makes the processor more tolerant of branch mispredicts. This is a bit less obvious, because the SMT processor does not naturally schedule around mispredicts the way it schedules around cache misses. However, the increased tolerance to branch mispredicts comes from two sources. First, the processor is typically following each (per thread) path less aggressively, and fewer fetched or executed instructions are lost when a branch mispredict is discovered—this is a result of the fact that we are relying less on speculation to provide the necessary instruction level parallelism, and instead relying on thread-level parallelism to supply the ILP. Second, even when one thread is servicing a branch mispredict and not making forward progress (or even while it is fetching down the wrong path before the mispredict is discovered), other threads in the processor typically are making forward progress.

It is shown that the latter effect (tolerance of mispredicts) is more pronounced than the former (increased mispredicts due to aliasing between threads), as overall an SMT processor is less sensitive to the quality of the predictor than single-thread processors [Ramsay et al., 2003, Serrano, 1994, Tullsen et al., 1996].

While a completely unmodified branch predictor, then, may still work well on a multithreaded processor, the branch target buffer will have to be modified slightly to include thread ID or process ID bits per entry, so that we do not produce phantom branch predictions for the wrong thread.

Hily and Seznec [1996] examine three different predictors on an SMT architecture. They show that the impact of sharing varies among the predictors, but in general the negative interference of a multiprogrammed workload was relatively minor. For parallel workloads, there was gain from the positive interference between threads, but again it varied (*gshare* showed the most gain) and the gains were also relatively small. Ramsay et al. [2003] confirm that shared prediction tables, even with a full compliment of threads, significantly outperform partitioned tables.

[Choi et al., 2008] examine, specifically, architectures or programming models that create many short threads (e.g., speculative multithreading, transactional memory). They show that conventional branch predictors that incorporate global history can be nearly useless for threads that execute only hundreds of instructions.

# 6.3   MANAGING CONTENTION THROUGH THE FETCH UNIT

The fetch unit is the gateway to the processor core, and in many ways the most important shared resource to control. In fact, it has been shown that if you want to control the flow of instructions that go through the execution units, it is more effective to do that by controlling the fetch streams than by manipulating the issue logic [Tullsen et al., 1996]. Most of the research into multithreaded fetch has been done in the context of SMT processors. That is due in large part to the fact that SMT is typically paired with out-of-order instruction execution, which is particularly vulnerable to stalled threads monopolizing instruction window resources. It is well documented [Cristal et al., 2004, Martínez et al., 2002, Wall, 1991] that processor performance scales with scheduling window size (e.g., the instruction queue size); thus, one thread that is stalled holding half of the window resources can significantly degrade the ability of other threads to exploit instruction level parallelism.

There are generally three goals, then, of an SMT instruction fetch mechanism. The first is to achieve high fetch throughput so that the fetch unit does not become a bottleneck—this is more of an issue with multithreaded processors because the execution throughput is much higher, making it more difficult for the fetch unit to stay ahead of execution. The second and third goals are closely related—the second goal is to maximize parallelism within the instruction scheduling window, and the third is to minimize the scheduling and execution resources held by stalled threads.

While sharing the fetch unit among multiple units creates a challenge, it is also a significant opportunity. We can exploit the high level of competition for the fetch unit in two ways not possible with single-threaded processors: (1) the fetch unit can fetch from multiple threads at once, increasing our utilization of the fetch bandwidth, and (2) it can be selective about which thread or threads to fetch from. Because not all threads would provide equally useful instructions in a particular cycle, an SMT processor can benefit by fetching from the thread(s) that will provide the best instructions.

The first opportunity was shown to be important in the context of 8-wide superscalar cores, where it was difficult to fill the fetch bandwidth because basic blocks (which can be fetched consecutively with no intervening control flow) are typically smaller than eight instructions. Thus, fetching two basic blocks from two threads makes it much more likely that we fetch 8 instructions. However, in the context of modern 4-wide superscalar processors, it is less clear that it is necessary to fetch multiple threads per cycle.

The second advantage, the ability to be selective about which threads to fetch, is absolutely critical. There are two elements that must be present for a fetch policy to exploit this selectivity. First, there must be diversity in the quality of instructions to be fetched (e.g., one thread will bring in instructions ready to execute, while another will bring in stalled instructions). This is generally true—even for symmetric code, asymmetry in the cycle-by-cycle stall behavior of threads will create this diversity. The second element is less well understood – the threads that are not fetched must improve over time. If the second element is not true, we only delay the inevitable (fetching of poor instructions). Fortunately, the second is also true in many cases. Not fetching a stalled thread eventually allows the stall to clear, making that thread a good candidate for fetching. Not fetching a thread because it is too speculative (instructions unlikely to commit) allows branches to resolve and makes it less speculative.

Tullsen et al. [1996] examined several policies for selecting threads for fetch. They included BRCOUNT (fetch from the least speculative threads by counting unresolved branches), MISS-COUNT (fetch from the threads with fewest outstanding cache misses), and ICOUNT (fetch from the threads with the fewest instructions in, or heading toward, the instruction queue). ICOUNT significantly outperforms the others because (1) it maximizes the parallelism in the queue by always ensuring the most even mix of instructions from threads, and (2) it adapts to *all* sources of stalls, rather than identifying particular stall effects. Any stall (whether from the cache, FP operations, long dependence chains, TLB miss, etc.) will result in instructions accumulating in the queue, immediately making that thread low priority for fetch.

Tullsen and Brown [2001] showed that as memory latencies crept up into the hundreds of cycles, ICOUNT became less robust in handling long latency operations. When a thread is stalled for that long, two things can happen. First, even very rare opportunities for fetch can still allow that thread to eventually fill the shared structures (instruction queues, reorder buffer) and prevent others from making forward progress. Second, even if the thread only holds its share of the resources it still adversely effects other threads. For example, with two threads running, ICOUNT will allow the stalled thread to hold half the instruction queue, meaning the single non-stalled thread is restricted to half the machine even though it is the only thread actively running for a long period. Thus, they supplement ICOUNT with more drastic measures when a long latency miss is detected. The most effective is FLUSH, which actually flushes all instructions from a stalled thread, freeing resources for other threads, and requiring the stalled thread to refetch those instructions when the load miss returns.

Cazorla et al. [2004a] introduce DCRA (Dynamically Controlled Resource Allocation), which monitors a wider set of resources (each instruction queue is considered separately, as well as integer and floating point registers) than the ICOUNT mechanism (ICOUNT primarily monitors the IQs in aggregate), ensuring that no thread exceeds its allocation of any of those resources. Additionally, they distinguish between memory-intensive and CPU-intensive threads—in contrast to the prior approaches, which strive to reduce resources held by frequently stalled threads, DCRA assigns *more* resources to the memory-intensive threads, providing them more opportunity to find instruction and memory parallelism within those threads.

El-Moursy and Albonesi [2003] show that predicting load misses, when successful, can enable better fetch policies than detecting load misses (or backed up instructions from a load miss, as in ICOUNT), because it minimizes the number of stalled instructions that get brought into the queues. That is, prediction (when correct) allows them to be proactive rather than reactive, allowing fewer stalled instructions into the window before the stall is detected. Eyerman and Eeckhout [2007] improve upon FLUSH, and better target the MLP (memory level parallelism) that DCRA strove to address. In this case, they explicitly predict the MLP available at each load. For a load miss with no expected MLP, there is no gain to continuing to fetch that thread, but when there is MLP, they execute just far enough to exploit that MLP (and get subsequent cache miss events into the memory hierarchy) before applying the FLUSH mechanism.

Choi and Yeung [2006] apply a dynamic hill-climbing approach to learn the best allocation of resources to threads. They sample different partitionings of resources (near the current setting) and use the resulting performance differences to establish the shape of the performance hill. Like DCRA, they monitor a number of resources, but again they control access to resources via the fetch unit, by limiting the fetch of threads that have exceeded their share of a particular resource.

## 6.4   MANAGING REGISTER FILES

One of the most critical differences between a multithreaded architecture and a single-thread architecture, for a processor architect, is the need to replicate architectural register state. Increasing the size of a physical register file can be difficult because the large number of read and write ports required by many modern superscalar architectures already makes register file access a potential critical timing path. There have been a number of works, then, that address the size of the register file. Most of these apply to both multithreaded and single-thread architectures, but the problem is heightened in the multithreaded case.

For machines that lack register renaming, it is difficult to do much besides just replicate the register structure. In more aggressive machines that have register renaming, register file pressure is a combination of the permanent architectural state and the demands of in-flight instructions. The latter, in particular, creates many opportunities for optimization.

A number of proposals note that renamed registers are not released until they are overwritten; however, a register may have been dead for a long time before it is overwritten. The time between the last read of a register and its redefinition represents unnecessary storage of the value. These works use software [Jones et al., 2005, Lo et al., 1999] or hardware [Monreal et al., 2002] approaches to identify the last access to a register, allowing it to be freed early. Wallace and Bagherzadeh [1996] advocate doing register renaming later in the pipeline so that stalled instructions do not consume a register before they are ready to produce a value.

Other proposals [Borch et al., 2002, Butts and Sohi, 2004, Cruz et al., 2000, Oehmke et al., 2005] suggest some kind of register hierarchy, where the register file that sits next to the functional units is only a cache of a larger register file.

Physical register inlining [Lipasti et al., 2004] notes that narrow registers (those in which a small number of significant bits are being used) could be placed in the register map (using bits that would otherwise be used to hold a pointer to a physical register), eliminating a need for a register to hold that value. Register packing [Ergin et al., 2004], conversely, packs multiple narrow width values into a single register. Others combine registers that have the same value into a single mapping [Jourdan et al., 1998].

Mini-threads [Redstone et al., 2003] create a new class of thread that allocates a new context but shares the register state of a parent thread. This allows us to further increase the number of threads supported without growing the critical register file. Software can choose to partition the use of the available register state or use shared registers for communication.

## 6.5   OPERATING SYSTEM THREAD SCHEDULING

Multithreading processors can present a challenge for the Operating System, because they maintain characteristics of both a multiprocessor and a uniprocessor, but are neither. The changes required to port an operating system and make it work are minor [Redstone et al., 2000], because the multithreaded processor can always be treated like a multiprocessor, where each hardware context is a separately scheduled processor "core." However, that approach does not provide the optimal performance, as it ignores the sensitivity of individual threads to the execution characteristics of threads co-scheduled onto the same core. While multiprocessors, particularly chip multiprocessors, also experience the performance effects of resource sharing, it does not approach that of multithreading cores, which might share L1 caches, fetch bandwidth, branch predictors, scheduling windows, ALUs, cache/memory bandwidth, etc.

Snavely and Tullsen [2000] demonstrate that system performance can be highly sensitive to the co-scheduling of threads in a system where there are more threads to schedule than hardware contexts. Assuming a diverse set of jobs available which are potentially bottlenecked on different parts of the processor, co-scheduling jobs that do not experience the same bottleneck resource tends to maximize global performance; for example, scheduling an integer-heavy application with a floating point application is likely to perform better than scheduling two floating-point benchmarks together. They identify performance counter statistics that indicate when threads are cooperating (or not cooperating), and seek schedules that equalize those statistics across all co-scheduled sets of jobs. They call this schedule *symbiotic jobscheduling*. Eyerman and Eeckhout [2010] eliminate the need to do extensive sampling to identify symbiosis. Instead, they use careful performance monitoring to create a model of thread performance, allowing them to model, or predict, whether a group of threads will demonstrate symbiotic behavior.

Snavely et al. [2002] extend the symbiotic jobscheduling work to account for co-scheduled threads with different priorities. Dorai and Yeung [2002] also explore architectural support to enable a low-priority background thread to maximize throughput while minimizing its impact on a high-priority foreground thread.

Fedorova et al. [2005] examine the case of a chip multithreaded (CMP of multithreaded cores) architecture, seeking to identify coschedules of threads that all coexist well in the shared L2 cache. Devuyst et al. [2006] also looks at OS scheduling for a chip multithreaded architecture. In particular, they argue for a scheduler that considers unbalanced schedules (e.g., three threads on one core, one on another core). Unbalanced schedules better facilitate the proper allocation of resources to threads with diverse execution demands.

Heat and Run [Powell et al., 2004] examines scheduling for thermal management. Threads with complementary resource needs can improve thermal behavior when co-scheduled. For example, two floating-point intensive applications co-schedule may create a hot spot, while co-scheduling a floating-point intensive and an integer-intensive may not. Additionally, multithreaded processors on a CMP give the system greater control over power and thermal hot spot management, as the power dissipation rises with the number of threads. Threads can be migrated away from hot cores as hot spots arise.

Other work has looked at the specific problem of scheduling to minimize interference in the cache hierarchy. Kihm et al. [2005] show that the OS can schedule more effectively with finer-grain performance counters (that capture misses per cache region rather than the cache as a whole). This allows them to exploit applications that utilize the cache unevenly and schedule them opportunistically with others that do not stress the same region.

## 6.6   COMPILING FOR MULTITHREADED PROCESSORS

On a uniprocessor, or even a multiprocessor, the processor core that the compiler generates code for is static. However, on a multithreaded processor, the processor "core" that a single thread executes on is dynamic, changing every cycle based on the demands of co-resident threads. The number of functional units, instruction window size, cache space, etc. available to one thread depends on the resource demands of the other threads. The compiler, then, cannot precisely compile for the resources available, but it can take steps to minimize performance-reducing conflicts between threads.

Lo et al. [1997] identify several tradeoffs that change when compiling for an SMT processor. Traditional parallel compilation seeks to partition memory between threads to minimize false sharing, typically preferring *blocked* allocation of loop iterations to threads (where iterations 0 to n go to thread 0, n+1 to 2n to thread 1, etc.). But on an SMT processor, this actually maximizes pressure on the cache and TLB. Instead, using cyclic partitioning of iterations to threads (iterations 0, t, 2t, ... to thread 0, iterations 1, t+1, 2t+1 to thread 1) can provide superior SMT performance, allowing threads to benefit from sharing across cache lines and across TLB entries—false sharing between threads is typically positive with multithreading. Aggressive software speculation (e.g., trace scheduling) occupies unused resources on a single-threaded processor, but can hurt performance on an SMT processor running several threads. In fact, any optimization that seeks to achieve throughput via increased instruction count (e.g., predicated execution, software pipelining) needs to be re-examined in a multithreaded environment—the optimizations may still be worthwhile, but the best settings will likely change.

Cache optimizations, in general, should not assume they own the entire cache. Compiler optimizations for instruction cache performance [Kumar and Tullsen, 2002] and data cache performance [Sarkar and Tullsen, 2008] can enable better sharing of the cache space. If the applications that are likely to be co-scheduled are known a priori, they can be cooperatively laid out in memory—high-frequency procedures or data objects, whether from the same thread or different threads, can be laid out so that cache conflicts are minimized. If the threads that will be co-scheduled are not known a priori, all applications can be laid out so that they pressure the cache in an unbalanced manner, then the OS can ensure that their virtual-to-physical mappings cause the high-pressure regions to be offset in the cache. Other works also emphasize that applications should avoid using the whole cache; for example, tiling for a portion of the cache instead of the entire capacity can allow multiple threads to use the cache without thrashing [Nikolopoulos, 2003].

# 6.7    MULTITHREADED PROCESSOR SYNCHRONIZATION

In designing synchronization and communication mechanisms for multithreaded processors, two architectural differences vs non-multithreaded processors are particularly relevant—the sharing of core execution resources and the sharing of the lowest levels of the cache hierarchy. Thus, we will establish two goals for multithreaded communication. First and foremost, a thread stalled on synchronization should not slow down other threads. Second, synchronization that takes place between two threads on the same core should not leave the core (that is, not cause potential cache and interconnect interference outside the core).

The first goal simply means that a stalled thread (e.g., one that fails to acquire a lock or is waiting at a barrier) should never spin. Single-threaded cores frequently use spinning in these situations because the core is not being used for any other purpose and spinning allows the processor to react quickly when the lock becomes available. They only release the core (typically after some spinning) when there is reason to believe they will continue to spin long enough to justify the cost of a software context switch. A spinning thread in a single-thread core only impacts other active threads if it causes coherence actions—this is why those machines might use (at least) test-and-test-and-set rather than test-and-set approaches for spin locks [Anderson, 1990, Rudolph and Segall, 1984]. However, on a multithreaded processor, spinning uses resources that would otherwise be available to other threads. For FGMT and SMT, it uses fetch bandwidth, execution bandwidth, instruction queue space, cache bandwidth, etc.

HEP [Smith, 1982] and the Tera MTA [Alverson et al., 1990] both used full-empty bits on memory to enable efficient and spin-free synchronization. Each word has a full-empty bit that can either be set or conditionally used on each memory access. For example, you write a word and set the bit to full or you can read a word if full. On a conventional processor, one might do producer-consumer communication by having the consumer thread attempt to acquire a lock held by the producer. The consumer would write data to memory then release the lock, after which the consumer would acquire the lock and then read memory. With full empty bits, the consumer simply does a read-if-full and the producer does a write-set-full. In the HEP and Tera, a read-if-full (for example) that does not succeed right away (e.g., the word is empty) just looks like a load that takes a very long time to return from memory. Because those machines typically do not allow same-thread instructions in the pipeline at once, a thread stalled for synchronization consumes no pipeline resources. While both machines support full-empty bits on all memory locations, HEP also supports full-empty bits on registers.

Alewife [Agarwal et al., 1995] also supports full-empty bits on memory, as does the M-machine; however, the M-Machine follows the model of HEP and enables fast explicit synchronization between co-scheduled threads by supporting full-empty bits on registers (that are shared between threads) [Keckler et al., 1998].

The lock box [Tullsen et al., 1999] addresses the second goal. It allows software to use the same acquire lock and release lock semantics familiar to conventional systems, but releases and locks are caught and managed in hardware by the lock box. In this way, instead of an acquire that repeatedly fails, software will see an acquire that succeeds, but with a long latency. Thus, even code that is written to spin will not spin. Because the state of all threads waiting on locks is available to the hardware, a released lock can be delivered to a waiting thread without needing to update memory state in any way—thus requiring no coherence or even cache activity.

## 6.8  SECURITY ON MULTITHREADED SYSTEMS

Multithreading processors, especially SMT, raise some new security concerns. The more two threads conflict and interfere with each other, the more one thread can learn about the other. A side channel attack is one in which no explicit data passes between a secure program and an observer/attacker, but the secure program leaks information through the physical system on which it runs. This information can be timing, power, electromagnetic signatures, sound, etc.

Multithreading processors, then, are of concern because the timing (in particular) of one thread is impacted by the execution bandwidth, branch patterns, and memory access patterns of a co-resident thread. Aciicmez and Seifert [2007] examine the specific question of whether an SMT processor (in particular, the Pentium 4) is more vulnerable than a non-multithreaded processor. They contend that while SMT processors are vulnerable to data cache attacks, instruction cache attacks, and branch predictor attacks, these all attack persistent state which can be accessed even on non-multithreaded cores. They show that they can successfully launch a timing attack on a shared multiply functional unit to reveal details about an RSA process, demonstrating an attack that is specific to a multithreaded processor.

Demme et al. [2012] seek to quantify vulnerability to side-channel attacks in a generic way, using correlation between a victim's execution and what another program can observe. They introduce the *side-channel vulnerability factor*. They note that disabling SMT for secure processes can foil SMT-specific attacks, and statically partitioned resources can limit vulnerability. They show that SMT does increase vulnerability—much of the vulnerability comes from the shared caches, but they show additional vulnerability from the pipeline sharing, as well.

CHAPTER 7

# New Opportunities for Multithreaded Processors

The previous Chapter deals with a variety of contention issues arising from the low-level sharing of processor resources. However, that high level of sharing also creates opportunities not available to non-multithreaded processors.

The first opportunity takes advantage, or even creates, positive interference, allowing one thread to positively impact the performance of another. Specifically, we examine a number of mechanisms that all fall under the general heading of helper threads in Section 7.1.

Additionally, multithreading provides cheap parallelism. While this has general performance implications covered throughout this book, it enables other opportunities as well. As specific examples, we will examine SMT-based fault tolerance in Section 7.2 and speculative multithreading in Section 7.3. The cost of implementing redundancy for fault tolerance or speculative multithreading (or helper threads, for that matter) may be prohibitively high on a conventional multiprocessor. However, if we instead host the same techniques on a multithreaded processor, the incremental hardware cost is quite low (only the cost of a 2nd hardware context). So an architectural technique that cannot be justified when it requires two cores may be perfectly reasonable on an SMT core when the incremental cost of the additional "core" is small.

Last, we will examine the energy efficiency advantages created by the high levels of pipeline utilization enabled by multithreading in Section 7.4. There is always a power (e.g., static leakage) and energy (increased runtime) cost to leaving execution resources idle, something we expect to do much less of on a multithreaded processor.

## 7.1 HELPER THREADS AND NON-TRADITIONAL PARALLELISM

Throughout the previous chapter we saw that due to the high incidence of sharing, multithreaded processors are vulnerable to negative interference between co-located threads. However, that sharing can also be put to our advantage to create positive interference. This can happen with traditional parallel programs, where communicating data goes through the shared L1 cache rather than more distant levels of the hierarchy. This section, though, deals with threads created for the sole purpose of positively influencing other threads. These threads typically offload no useful computation from the main thread, but in most cases compute or acquire some useful state that enable the main thread to run faster. As a result, these threads are often called *helper threads*. Because this mechanism enables parallel speedup of serial code without partitioning or distributing the original computation, as is done with traditional parallelism, it is sometimes called *non-traditional parallelism*.

Helper threads have been proposed to provide several types of assistance to the main thread. The most common application is precomputation of addresses the main thread is going to load in the future, so that they can be prefetched into the data cache [Collins et al., 2001a,b, Kim and Yeung, 2002, Luk, 2001, Zhang et al., 2005, Zilles and Sohi, 2001]. However, helper threads could potentially also be used to precompute future branch outcomes, compute future instruction fetch addresses to support instruction cache prefetch, and to dynamically recompile the running code to customize it to the runtime behavior of the main thread.

Cache prefetching helper threads are particularly attractive, because they require no architectural changes to a multithreaded processor as long as the contexts share the cache. In addition, they provide parallelism without the concerns of data races and correctness—all the helper thread does is load data into the cache. If the values change after prefetch (e.g., by another thread), coherence still protects the data; and if the helper thread prefetches incorrect addresses, it only affects performance, not correctness. It has been shown that in many cases the majority of expensive cache misses are generated by a small number of static loads [Collins et al., 2001b]—this argues for a heavyweight mechanism (such as helper threads) to address those static loads.

In many cases, the code for prefetching helper threads is distilled from the original code. This is done by copying the original code, then eliminating any computation that does not produce memory addresses for the targeted loads. Speculating on control flow (e.g., eliminating unlikely branch paths) can significantly reduce the size of the distilled code (eliminating both the branch and the code that computes the condition). It is critical that the majority of the code be eliminated because the prefetching thread must run fast enough to stay ahead of the main thread, even when it is incurring misses and the main thread is not. Whenever possible, the helper thread should issue prefetch instructions rather than loads – in this way, if the control flow speculation is incorrect, invalid addresses will not cause exceptions. But this is not possible for chained loads, where the loaded value is needed to compute the next address. Chained loads (e.g., pointer chasing through a linked list or a tree) are a particular challenge to helper thread prefetchers (as well as traditional prefetch mechanisms). This is because there is no memory level parallelism and all memory accesses are serial, making it difficult for the helper thread to run faster than the main thread unless there is significant main thread computation that is not part of the address calculation.

Helper threads, particularly those distilled from the main code, can be generated in a number of ways. They can be written by the programmer, generated by the compiler [Kim and Yeung, 2002], generated in hardware [Collins et al., 2001a], or generated dynamically by other helper threads [Zhang et al., 2007].

Several proposals precompute branch outcomes [Chappell et al., 1999, Zilles and Sohi, 2001] (e.g., via a distilled version of the main thread). This requires hardware support to communicate these outcomes between hardware contexts. There are challenges in precisely matching these precomputed outcomes with the right branches (and especially the right dynamic instance of the static branch).

Helper threads have also been proposed to support speculative multithreading processors by precomputing dependent live-in data in lieu of speculating on the values [Madriles et al., 2008]. Aamodt et al. [2004] use helper threads to precompute and prefetch instruction cache accesses. In event-driven compilation [Zhang et al., 2005, 2007], helper threads are spawned when hardware performance monitors identify interesting runtime behavior, generating a lightweight compiler thread that modifies and adapts the executing code, then exits.

Chappell et al. [1999] introduce helper threading in simultaneous subordinate microthreading (SSMT), an architecture with specialized lightweight threads that accelerate the main thread. They discuss cache and prefetch opportunities, but demonstrate helper threads that precompute difficult branches. The first proposals that demonstrated prefetch helper threads on a simultaneous multithreading architecture were Collins et al. [2001b], Zilles and Sohi [2001], and Luk [2001], which all appeared at the same conference. The first two generate helper threads by distilling code from the main thread, the other uses programmer-generated helper threads customized to each application. The first also examines branch precomputation.

## 7.2 FAULT TOLERANCE

Multiprocessors have long been recognized as an attractive platform for fault tolerant solutions, because they provide natural redundancy. That redundancy can be exploited by performing computation on two or more processors or cores and comparing the outcomes. In particular, multiprocessors provide *space redundancy*, allowing redundant operations to take place in parallel.

A single multithreaded core does not provide the same space redundancy, as multiple virtual cores (contexts) share a single physical core. However, multithreaded processors can still be a highly attractive platform for fault tolerant execution, even compared to CMPs, because they provide redundancy at a much lower cost than fully replicated hardware. Fault-tolerant multithreaded proposals typically exploit *time redundancy* rather than space redundancy, although the latter is not completely unavailable (e.g., two threads could use different integer units of a superscalar SMT core).

Time redundancy can detect *transient faults* (faults that cause a combinational or state element to produce a wrong value once), but will not be effective against *hard faults* (a permanent fault in a logic element that will cause repeatable errors). Transient faults, however, are particularly important, as (1) they are expected to grow significantly with process shrinkage and aggressive voltage scaling, and (2) they cannot be detected by built-in self test (BIST) and related techniques.

The downside of exploiting time redundancy via multithreading is that the redundant threads compete for the same execution resources, which can slow down both threads (or all threads, but our discussion will assume two threads for simplicity). In some cases, neither thread uses the full bandwidth of the machine, and the two threads co-exist nicely. This is the same phenomenon we exploit in normal multithreading, but redundant threads have an advantage because they do not experience negative competition for all resources—they tend to positively interfere in the branch predictor and may share the same working set in the cache (depending on whether the memory image is unified or replicated).

This positive interference can be enhanced by intentionally creating a minimum time differential between the two threads. This time differential allows branch directions and load/store addresses to be resolved in one thread before they are needed in the other thread. In this mode, one thread becomes the computation thread and the trailing thread becomes the checker. The computation thread then becomes a highly accurate predictor of the checker thread—branch outcomes, load addresses, and even data values (the result of not-yet-executed instructions) can be predicted perfectly in the absence of faults. This does not eliminate the ability to detect differences in the instruction stream, because the data from the leading thread is inserted into the trailing thread as predictions—differences are simply detected as mispredictions in the trailing thread. In many cases the same misprediction detection that already exists in a speculative architecture is now used in this case to detect errors. These predictions provided by the leading thread not only speed up the checker thread, but also reduce its hardware footprint—by never mis-speculating, it uses less fetch and execution bandwidth.

AR-SMT [Rotenberg, 1999] identified the potential for SMT processors to exploit time redundancy and, in particular, to use the leading thread (active thread, or A-Thread) as a predictor for the trailing thread (redundant thread or R-Thread). The SRT (simultaneous and redundantly threaded) processor [Reinhardt and Mukherjee, 2000] narrows the sphere of replication to just the processor hardware, assuming the memory hierarchy can be protected by ECC. By doing this, there is no extra pressure on the caches, and error detection can be employed at minimal cost by just checking stores and uncached loads. The Load Value Queue enables a single cache/memory access to deliver the result to both threads. They also employ fetch slack (enforcing a time differential between the two threads), to ensure that branch outcomes are available and cache misses have been resolved before the trailing thread needs them.

In these solutions, only the trailing thread is accelerated so the pair run no faster than a single program. The Slipstream processor [Sundaramoorthy et al., 2000] solves this by relaxing full redundancy. The trailing thread becomes the reference execution and the advance thread's primary purpose is to provide the predictions. Thus, the advance thread can employ aggressive speculation (skipping computation likely to not impact long-term program state), knowing it will be corrected by the trailing thread. Thus, the advance thread executes fewer instructions and runs faster, while the trailing thread runs faster due to the presence of predicted state.

## 7.3 SPECULATIVE MULTITHREADING ON MULTITHREADED PROCESSORS

Speculative multithreading [Marcuello et al., 1998, Sohi et al., 1998], also called thread level speculation, refers to processor architectures that speculatively parallelize serial code. If the regions executed in parallel are free of dependencies, the parallel execution succeeds; if a dependence violation is detected (e.g., a memory read in a younger thread followed by a memory write in an older thread), the younger thread may be squashed and restarted. This allows the parallelization of code not written to exploit parallelism, or which a compiler or programmer could not parallelize using conventional means because of dependence uncertainties.

Multithreading processors provide three advantages for speculative multithreading. First, as discussed at the beginning of this chapter, it provides cheap parallelism. Speculative multithreading techniques have struggled to deliver huge performance gains, even on simulated hardware, primarily because they target code for which traditional parallel techniques have failed. However, justifying the area and power of an SMT thread, for example, represents a much lower barrier than that of a CMP core. Second, speculative multithreading threads tend to communicate heavily because (1) they do tend to have more (true and false) dependencies, since the code was typically not compiled to separate the threads, and (2) spawning and committing threads requires additional communication. Speculative multithreading on a single core isolates that communication, rather than having it all go across the shared on-chip interconnect. Third, spawning new threads typically requires the copying of register state between contexts. On a CMP, this is an expensive operation, possibly taking 10s to 100s of cycles (in some architectures, that is longer than the typical speculative thread). On an SMT processor, this can be done with a simple copy of a context's register map, perhaps in a couple of cycles.

A number of research efforts have looked at the implementation of speculative multithreading on multithreading (typically SMT) cores. The Dynamic Multithreading Processor [Akkary and Driscoll, 1998] seeks to maximize fetch bandwidth when executing a single program by dynamically breaking execution up into threads. It identifies future execution paths in two ways. First, it identifies loop-back branches (speculatively spawning the fall-through loop exit path) and procedure calls (speculatively spawning the post-return path). All threads share pipeline resources as in other SMT designs. They add trace buffers to buffer speculative non-committed state, logic to speculate on register data values and register and memory dependencies, and of course logic to detect misspeculation.

The Data Speculative Multithreaded Architecture [Marcuello and González, 1999] includes many of the same features. They speculate on future loop iterations, and exploit the shared nature of the SMT fetch unit to achieve a high fetch rate. They fetch from a single PC, but each context (executing a different iteration) renames the fetched instructions separately, thus multiplying the effective fetch rate.

The Implicitly-Multithreaded Processor [Park et al., 2003] also exploits SMT's shared register state for efficient speculative multithreading. It assumes the compiler identifies thread spawn points and inter-thread register communication. They avoid adding complexity to key pipeline structures by maintaining the in-order nature of structures such as the active list and the load-store queue, despite out-of-order fetch of instructions from different threads. They do this with the dynamic resource predictor which predicts the exact resources each thread will need, and reserves space between threads, as needed. They also modify fetch policies to account for the relative priority of more speculative or less speculative threads.

Packirisamy et al. [2008] specifically examine the question of whether an SMT processor or a CMP make a better host for speculative multithreading. Because they assume an equal-area comparison, the CMP configuration uses simpler processors and therefore consumes less power. However, due to better communication and the gains from threads sharing the L1 cache, the SMT configuration achieves 21% better performance and 26% better $energy \times delay^2$. Warg and Stenstrom [2006] conduct a similar study, looking only at performance. They show that due to significantly lower thread management and communication costs, SMT-only speculative threading outperforms CMP implementations. In fact, even a 2-thread SMT configuration is competitive with a 4-core CMP.

## 7.4   ENERGY AND POWER

Multithreaded processors are naturally energy efficient. They can be designed to use less power than a single-threaded core, in which case they combine the potential for high throughput with low power, maximizing energy efficiency. But even in the case where we make no special allowances for power, as in the high-performance SMT case, the multithreaded processor still has the potential to be significantly more energy efficient than the same core without multithreading.

Chapter 4 describes a fine-grain multithreaded processor core that enables the elimination of several power-hungry structures, including inter-instruction dependence checking, bypass logic, and branch prediction. In addition, this architecture wastes no energy on speculative execution. This is not the only way to design a fine-grain multithreaded processor, but in the presence of abundant thread-level parallelism, this is an attractive design point which can provide high throughput without those structures. This creates an extremely energy efficient architecture.

But energy efficiency is not exclusively the domain of a conservatively designed fine-grain processor. In the initial work on simultaneous multithreading architectures and power, Seng et al. [2000] describe several advantages that make SMT architectures energy efficient, compared to a similar pipeline without multithreaded support, even though the SMT architecture tends to use more power (due in large part to increased activity due to higher instruction throughput). First, the SMT architecture relies much less heavily on speculation to achieve performance. Second, multithreaded architectures amortize the power that is consumed much more effectively. Underutilized resources waste power, and multithreading maximizes utilization. Between static leakage and dynamic power consumed by structures that will be active whether single-thread or multithreaded, processors consume a certain level of baseline power when active—this power is amortized over a higher number of executed instructions when multithreading, minimizing energy per useful instruction. Third, multithreaded architectures can be even further modified to decrease speculation, often with minimal impact on performance. Additionally, a multithreaded processor can be made to degrade gracefully (simply decreasing the thread count) during power or thermal emergencies.

Researchers from IBM, Virginia, and Harvard [Li et al., 2004] created a detailed power model of a Power4-like SMT processor core to evaluate the energy efficiency of simultaneous multithreading. They separated the two effects of (1) power for increased structures sizes, like register files and instruction queues, and (2) power consumed by increased activity due to higher instruction through-put. They did so under varying assumptions about the increase in size of the hardware structures. Using $energy \times delay^2$ as the energy efficiency metric, they show that SMT provides a 10-30% increase in energy efficiency; further, energy efficiency is maximized under fairly aggressive sizing of structures. That energy efficiency comes from improving performance more than we increase power. That work breaks the increased power down into the two categories above, for that same region where energy efficiency is maximized. They show that the total power increase incurred by SMT is about 24% of baseline power. Of that, about 15% comes from the increased structure sizes, and about 9% comes from increased utilization.

A number of studies on real processors have shown the increased energy efficiency of simultaneous multithreading, including on the Intel Pentium4 [Marr, 2008] and the Intel i7 [Schöne et al., 2011].

# CHAPTER 8

# Experimentation and Metrics

Multithreaded execution poses several challenges for measuring, simulating, and reporting performance, particularly when running a multiprogrammed workload in a simulated environment—e.g., one which measures instruction throughput over an interval rather than full runtimes. These issues are not isolated to hardware mulithreaded architectures, but are shared by any multiprocessor. However, most multiprocessor research, particularly before multicores, ran or simulated parallel benchmarks—with parallel benchmarks these issues are less of an issue, if at all. As a result, these issues were first exposed and explored at length in the multithreading research.

The two primary challenges faced by multiprogrammed experimentation are the *repeatability* problem (Figure 8.1) and the *measurement interval* problem (Figure 8.2). The repeatability problem stems from the fact that in a multiprogrammed experiment threads run independently, and thus the relative progress of each thread is not the same from experiment to experiment. We see in Figure 8.1, where each line represents the progress made by a particular thread in an experimental run, that the actual instructions run in experiment B are not the exact same instructions run in experiment A, even though there may be significant overlap.

This problem is exacerbated if there are phase changes during the measurements—see the red line in Figure 8.1, where the width of the line represents a change in raw IPC (instructions per cycle). We run a greater fraction of instructions from the high-IPC phase in experiment B than we did in experiment A.

One solution is to enforce that we run the exact same instructions in experiment B; unfortunately, that introduces end effects that are even more disruptive. Experiment A will show good performance because the threads are artificially balanced, while experiment B would have a tail where

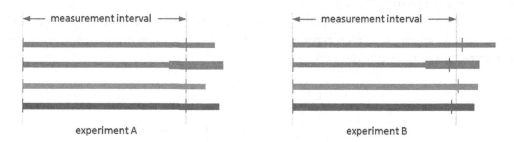

**Figure 8.1:** The repeatability problem in multithreaded simulation. Successive experiments do not run the same instructions for each benchmark.

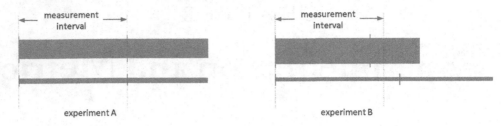

**Figure 8.2:** The measurement interval problem. By changing the architecture to increase IPC in the measurement interval, we see an artificial increase in performance, as we only postpone low-IPC jobs until after the measurement interval.

three threads, then two, then one thread is running alone at the end of the simulation. But that in no way indicates that experiment A represents a better architecture or configuration.

The measurement interval problem is also reflected in Figure 8.1, but is more directly illustrated in Figure 8.2. All experiments measure performance over a measurement interval, but we must keep in mind that measurement interval always represents only a portion of real execution. In this figure, we see the danger of using IPC as a metric for performance. Here, we have a high-IPC thread (blue) and a low-IPC thread. If we make an architectural change that favors the high-IPC thread, we will push more of the high-IPC thread into the measurement interval, thereby increasing overall IPC over the measurement interval. But we have not gained real performance, we have merely sacrificed performance outside the measurement interval to artificially boost performance inside the interval. Eventually, in a real system, we'll have to run all of the low-IPC threads that got delayed. In many cases, IPC will actually be *counter-indicative* of real performance gain, as in this figure where total execution time is elongated.

When we run a parallel workload, it is common to measure the time to execute from the beginning of a parallel region to the end. This sidesteps these problems because (1) by measuring this region we are either measuring the same instructions, or if not, we are measuring the same amount of progress through the program, and (2) we are measuring execution time directly, instead of a secondary measure like IPC.

These problems do not crop up in single-thread experiments, because (1) it is easy to run the exact same instructions in each experiment, and (2) IPC is actually a good proxy for a meaningful performance metric—execution time ($T_{ex}$). That is,

$$T_{ex} = \frac{instruction\_count * cycle\_time}{IPC} \tag{8.1}$$

Thus, as long as we are executing the same instructions and cycle time is unchanged:

$$IPC \propto \frac{1}{T_{ex}} \tag{8.2}$$

The problem is that with a multiprogrammed workload, aggregate IPC is not related in any useful way to the execution times of the constituent programs in the workload. In fact, it is easy to construct examples where aggregate IPC goes up while average execution time of the workload degrades. For that reason, aggregate IPC needs to be considered a completely unreliable indicator of performance, and should never be used as a primary metric in these types of experiments.

This problem was first addressed by Snavely and Tullsen [2000], who introduced *weighted speedup* as a performance metric. Originally intended to express the speedup of multithreaded execution over single-thread execution, weighted speedup (WS) was expressed as:

$$WS_{multithreading} = \sum_{i=1}^{n} \frac{IPC_{MT,i}}{IPC_{ST,i}} \tag{8.3}$$

where $IPC_{MT,i}$ is the IPC of thread $i$ when running multithreaded, and $IPC_{ST,i}$ is the IPC of thread $i$ when running alone. For example, if we run two threads together on a multi-threaded processor, and each runs at 70% of the speed they would run if they had the processor to themselves, we get a weighted speedup of 1.4. Weighted speedup, then, requires that we identify a baseline architecture, which in this case is single-thread execution. Weighted speedup was generalized in Tullsen and Brown [2001] to handle what would become the more common case, where the baseline architecture we were trying to improve was itself multithreaded.

$$WS = \frac{1}{n} \sum_{i=1}^{n} \frac{IPC_{opt,i}}{IPC_{base,i}} \tag{8.4}$$

In this definition of WS, they compare the IPC in an optimized (opt) configuration relative to some baseline (base). They introduce the $1/n$ factor in this context, where both the baseline and the optimized architecture are multithreaded, to get an average speedup (typically near one). Thus, if one thread experiences a 20% speedup and the second thread a 10% slowdown (.9 speedup), the weighted speedup would be 1.05.

This addresses the measurement interval problem with IPC because with weighted speedup, we value a 10% speedup equally on any thread, regardless of whether it is a high-IPC thread or a low-IPC thread—i.e., in Figure 8.2, speeding up the red thread has the same value as speeding up the blue thread. More concretely, speeding up the blue thread by 5% at the expense of slowing down the red thread by 10% would increase IPC but decrease weighted speedup.

This does not, however, completely address the repeatability problem, unless IPCs for the threads are reasonably constant across the measurement intervals. This is not always the case, especially since the single simpoints [Sherwood et al., 2002] used in many architectural studies to identify instruction regions for experimentation often contain phase changes. For this reason, later work [Tune et al., 2004] more carefully defined $IPC_{base,i}$ to be the IPC of the baseline configuration running the *exact* same instructions as executed by that thread in the optimized run. To enable that calculation, the baseline experiment must be run for a much longer period, with frequent capturing of the IPC of each thread, so that the baseline IPC can be calculated no matter how many instructions a thread runs in the optimized experiment. Thus, if a thread runs in one phase 80% of the time, and another 20% of the time, that will be reflected in both the baseline IPC and the optimized IPC used in the calculation of weighted speedup for that thread.

Weighted speedup is based on the premise that individual IPCs can give us insights about individual job runtimes that aggregate IPC obscures. If we ran a number of jobs to completion and calculated a common multiprogrammed performance metric, *average speedup*, it would simply be:

$$Average\ Speedup = \frac{1}{n} \sum_{i=1}^{n} \frac{T_{base,i}}{T_{opt,i}} = \frac{1}{n} \sum_{i=1}^{n} \frac{IPC_{opt,i}}{IPC_{base,i}} \tag{8.5}$$

The last equality holds as long as the cycle time is unchanged. Thus, weighted speedup simply calculates the expected average improvement in execution time, assuming that the ratio of IPCs over the measurement interval is reflective of the ratio of IPCs over the whole program.

As an alternative to weighted speedup, Luo et al. [2001] proposed harmonic mean speedup. Harmonic mean speedup (HMS) simply inverts the IPC ratio before summing, then inverts again to get a metric where higher is better. It is defined as:

$$HMS = \frac{n}{\sum_{i=1}^{n} \frac{IPC_{base,i}}{IPC_{opt,i}}} \tag{8.6}$$

Since both WS and HMS work with ratios (which are unitless), it is not clear whether arithmetic mean or harmonic mean is more appropriate to combine them [Smith, 1988]. HMS can be derived from average normalized response time in the same way that WS is derived from average speedup. HMS was proposed because it was felt to also incorporate fairness. That is, high speedups are damped, and low speedups are magnified with harmonic mean. Thus, two threads that get 10% speedup would achieve a higher HMS than the case where one thread gains 15% and another gains 5%. Vandierendonck and Seznec [2011] show that WS and HMS are highly correlated for a series of experiments, and conclude that HMS does not add significant information to WS. Eyerman and Eeckhout [2008] advocate the use of more system-level metrics for multi-threaded/multiprogram performance, namely average normalized turnaround time (ANTT) and system throughput (STP), but when again restricted to a finite measurement interval and IPC measurements, ANTT reduces to HMS and STP reduces to WS.

Other studies have attempted to explicitly measure fairness in a separate metric. Gabor et al. [2007] introduce a fairness metric based on the unevenness of the slowdowns experienced by different threads.

$$Fairness \; = \; \frac{min(\frac{IPC_{MT,i}}{IPC_{ST,i}})}{max(\frac{IPC_{MT,i}}{IPC_{ST,i}})} \tag{8.7}$$

This is again expressed with single-thread execution as the baseline, but again can be generalized to the ratios of an optimized system vs. a baseline system. With the fairness metric, if all threads slow down (or speed up) uniformly, the metric will be one. Any imbalance in the slowdowns, and the fairness will be below one. Vandierendonck and Seznec [2011] advocate *minimum fairness* which is based on the numerator of the above equation, the thread with the minimum speedup (maximum slowdown).

Another proposed fairness metric [Kim et al., 2004] uses differences between the relative slowdowns rather than ratios. E.g., for two threads i and j, comparing a baseline system to that system with some modification.

$$Fairness \; = \; \left| \frac{T_{mod,i}}{T_{base,i}} - \frac{T_{mod,j}}{T_{base,j}} \right| \; = \; \left| \frac{IPC_{base,i}}{IPC_{mod,i}} - \frac{IPC_{base,j}}{IPC_{mod,j}} \right| \tag{8.8}$$

For more than two threads, they take the sum of all pairs of threads.

When measuring performance on real multithreaded systems, we can directly measure execution runtimes, which eliminates many of the challenges of reporting simulated results. However, there still remain a number of repeatability and variability issues that are more challenging than single-threaded machines; these challenges have also been addressed in the literature [Cazorla et al., 2010, Tuck and Tullsen, 2003]. Besides the repeatability concerns of any real-machine experiments, multithreaded experiments have an additional degree of complexity. Thread A's performance may vary based on how it is coscheduled with B (e.g., whether A starts at the same time or when B is halfway done), because of how the different phases interact.

CHAPTER 9

# Implementations of Multithreaded Processors

This Chapter gives more detail on a number of important implementations of multithreaded processors. It takes advantage of terminology and concepts we have established in prior chapters, Chapter 2 in particular. We focus on commercial implementations and important academic machines that produced real hardware. For a more comprehensive survey of academic multithreaded processor proposals, see Ungerer et al. [2003].

## 9.1 EARLY MACHINES—DYSEAC AND THE LINCOLN TX-2

The DYSEAC computer [Leiner, 1954], which began operation in 1954, was developed as a successor to the 1950 SEAC, one of the first electronic programmable computers. Both were designed and built by the U.S. National Bureau of Standards. DYSEAC features a form of master-slave coarse-grain multithreading, allowing it to hide the latency of I/O operations. The machine allows the programmer to maintain two program counters. On an I/O operation of long or indeterminate latency (indicated by specifying a bit in the *input-output* instruction), the machine will begin fetching at the alternate program counter. When the I/O operation completes, control is immediately transferred back to the primary program counter. In this way, the machine can enable independent computation while waiting. The DYSEAC was the world's first portable computer, housed in the back of a large truck.

The MIT Lincoln Lab TX-2 also featured coarse-grain multithreading (they referred to it as a *multiple sequence* computer). Their motivation was to manage a wide range of I/O devices, all time-sharing the central computer, and executing at different priorities. The machine can support up to 33 program counters. Because of a shared temporary register, control can not be given up at every instruction. Thus, an instruction will have the *break* bit set if it is a legal place to give up execution. If a higher priority thread is waiting, the hardware then context switches to the new stream. If a thread wants to give up control of the central computer completely (e.g., to wait for an I/O operation to complete), it will execute the *dismiss* instruction.

## 9.2 CDC 6600

The CDC 6600 [Thornton, 1970], delivered in 1964, was composed of a high speed central processor and 10 peripheral processors to which I/O jobs could be offloaded. The central processor

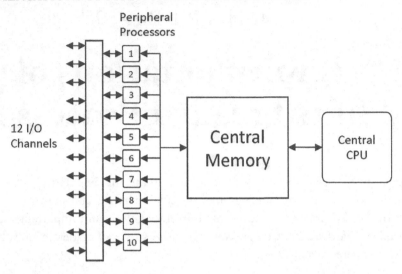

**Figure 9.1:** Architecture of the CDC 6600.

introduces scoreboarding to enable out-of-order execution, but for our purposes, the peripheral processors are more interesting. The peripheral processors (in aggregate) feature fine-grain multi-threading. A peripheral processor executes a different ISA than the central processor, supporting logical instructions, addition, subtraction, shift, and conditional jump. It can access both the central memory (60-bit words) and memory private to each peripheral processor (12-bit words).

To provide high throughput at a low cost, the 10 peripheral processors time-share a single execution unit, which runs at 10X the speed of a memory access (1000 ns). The state of each peripheral processor moves continuously through a barrel shifter, which brings the state (the hardware context) of one processor (thread) to the execution unit exactly once every ten cycles. This is facilitated by the fact that the state is extremely small—52 bits holds the program counter, accumulator (single register), status bits, etc.

The barrel shifter approach means that one thread can execute an instruction every 10 cycles, regardless of whether there are one or 10 active threads. Thus, every peripheral processor appears to be a single processor with execution and memory speed perfectly balanced, capable of executing an instruction (or in some cases a "step" of an instruction) every 1000 ns. In aggregate, it can execute an instruction every 100 ns with no memory stalls.

## 9.3    DENELCOR HEP

The Denelcor HEP [Smith, 1982] (Heterogeneous Element Processor) was an FGMT machine, architected by multithreading pioneer Burton Smith to compete with analog computers for executing non-linear ordinary differential equations. These algorithms exhibit fine-grain parallelism, but do

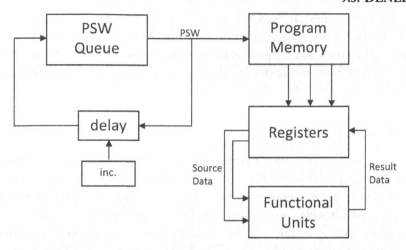

**Figure 9.2:** Architecture of the HEP Process Execution Modules, including the scheduling mechanism which utilizes the PSW queue. The next thread to execute an instruction is chosen each cycle out of the PSW queue, and the thread is not available to execute again until the PSW returns to the queue after circulating through a delay loop.

not map well to vector operations. Thus, the HEP integrates several data-flow type elements, such as support for many small threads and very fast communication and synchronization between threads. It was first delivered in 1981.

The HEP was designed to support up to 16 Process Execution Modules (PEMs) even though the largest delivered system had only four. The PEMs are connected to data memory and I/O over a high-speed packet switched network. Each PEM supports up to 128 independent instruction streams and 16 protection domains (distinct address spaces). Each PEM supports 2048 64-bit general purpose registers (and 4096 constant registers). Memory and register protection and isolation are implemented with base and limit registers for both registers and memory.

A single base and limit register per protection domain (for the general registers) means that threads in the same protection domain share not just memory but also registers—this requires unique synchronization solutions (described below) but allows extremely fast communication between co-operating threads. Because the state of a thread (including registers) is both much larger than the CDC 6600 and also not unique to a single thread, the barrel shifter approach was not an option. However, the HEP implements scheduling by sending a thread's PSW (process status word) through a timing loop (Figure 9.2) which guarantees that the thread's PSW does not return to the PSW pool to be rescheduled until the currently scheduled instruction completes—thus, the path through the timing loop depends on the expected latency of the instruction. Therefore, scheduling is not constrained to be round-robin, but still prevents instructions from the same thread from being in the pipeline at once.

The HEP associates hardware full/empty bits with each general register and memory location in the machine. For the registers, in particular, this not only enables extremely fast synchronization, but also allows threads to share registers without constant race conditions and conflicts. Full/empty bits, as described in Section 6.7, enable a diverse and powerful set of communication and synchronization paradigms.

## 9.4  HORIZON

Horizon [Kuehn and Smith, 1988, Thistle and Smith, 1988] never made it to manufacture, but this architecture included a few ideas that were later implemented in other commercial processors. The Horizon inherits many basic concepts from HEP, and is also a fine-grain multithreaded processor. The design has 64 bit wide instructions typically containing three VLIW operations: a memory operation, an arithmetic or logical operation, and a control operation which can process branches as well as some arithmetic and logic ops. The Horizon increased ILP without adding significant hardware complexity through *explicit-dependence lookahead*, replacing the usual register reservation scheme. Every instruction includes an explicit 3-bit unsigned integer (N), that indicates the number of instructions separating this instruction from the ones that might depend on it. Therefore, the subsequent N instructions could overlap with the current instruction. This allows the machine to maintain several attractive features of fine-grain multithreading (no bypass or hardware dependence checking), yet still expose per-thread instruction level parallelism.

## 9.5  DELCO TIO

The first FGMT system centered around real time constraints was the Advanced Timing Processor (later known as the Timer I/O) developed in 1984 by Mario Nemirovsky at Delco Electronics (a division of General Motors) [Nemirovsky and Sale, 1992a,b]. The Timer I/O (TIO) processor was widely used in the engine control module (ECM) in GM automobiles for many generations, starting in 1987.

The real time constraints in the ECM consisted of large numbers of hard deadline tasks (e.g., spark plug timings and fuel injector pulses) in addition to other functions. To handle these hard deadline tasks and the other functions, the TIO was implemented as a VLIW processor architected with two separate pools of threads—SIM (Single Instruction Machine) and MIM (Multiple Instruction Machine) executed in an interleaved fashion. While the SIM pool consisted of 32 independent threads, the MIM pool consisted of 8 independent threads. Each cycle an instruction from the active MIM thread would be chosen to execute followed by an instruction from one of the (active) 32 SIM threads in the next cycle. Hence, an SIM thread could only execute one instruction every 64 cycles whilst a MIM thread would execute an instruction every other cycle until completion where it would then choose another thread (of the 7 remaining) to run until completion. This architecture allows both high priority and low priority threads to share the processor, yet still have deterministic realtime runtimes.

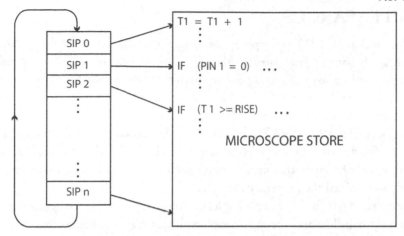

T1 = T1 + 1
    ⋮

IF   (PIN 1 = 0)  ...
    ⋮

IF   (T 1 >= RISE)  ...
    ⋮

MICROSCOPE STORE

SIP 0
SIP 1
SIP 2
⋮
⋮
SIP n

## Single Instruction Machine (SIM)

**Figure 9.3:** Delco TIO SIM.

## 9.6    TERA MTA

Burton Smith's Tera MTA architecture [Alverson et al., 1990] was based heavily on his prior two machines, the HEP and Horizon. It combines FGMT with 3-operation VLIW, uses full-empty bits on memory for synchronization and communication, and also includes explicit look-ahead set by software to enable per-thread ILP. The lookahead in this case indicates the number (up to eight) of subsequent instructions from the same stream that are independent of a memory operation. The goal of the FGMT bypass-free pipeline was to drive the clock rate up as high as possible for maximum performance.

One thing that separated the Tera architecture from contemporary designs was the absence of data caches. Latencies to memory are on the order of 140-160 cycles. These large latencies are tolerated by a combination of aggressive multithreading, explicit lookahead, and a high-bandwidth interconnection network. The MTA keeps the context of up to 128 threads in the processor in hardware. Each stream includes a program counter and a set of 32 registers; a stream holds the context for one thread. This processor also uses a timing loop (21 cycles because of the deep pipeline) to separate instructions from the same thread, so a minimum of 21 streams are required to keep a processor 100% utilized even if no instructions reference memory. A single thread running on the MTA, then, executes at less than 5% of the available pipeline bandwidth. It is optimized for many-thread throughput, not single-thread throughput.

## 9.7 MIT SPARCLE

Sparcle [Agarwal et al., 1993] is the processor designed for the MIT Alewife research multiprocessor. The design goals for the processor of Alewife included tolerance of long memory latencies and fast message handling. By supporting coarse-grain multithreading, they quickly switch contexts on cache misses to hide those latencies, and are able to rapidly start up a message handler at any time without software switching overheads. This processor design, from MIT, Sun, and LSI Logic, was a modification of an actual SPARC processor implementation. As previously noted, the register requirements for a coarse-grain processor are very similar to that for register windows—a large partitioned register file, with the pipeline never needing access to more than one partition at a time. This greatly minimized the necessary changes.

Sparcle supports four hardware contexts—three for user-level threads, and one permanently devoted to trap handlers and message handlers. It can do a context switch in 14 cycles—three to detect a cache miss, three to flush the pipeline and save the program counter, and the remaining to move non-register state in and out of the processor and restart the pipeline with a fetch at the new program counter. Alewife supports full/empty bits in the processor, and this support must also be extended to Sparcle's cache controller.

## 9.8 DEC/COMPAQ ALPHA 21464

The Alpha 21464 [Diefendorff, 1999, Emer, 1999] was not strictly a commercial product, but it was formally announced and was well into the design stage when the product was canceled due to the acquisition of Compaq by HP (and the acquisition of the Alpha group by Intel). If not canceled, it would likely have been the first commercial general-purpose SMT processor. Chief architect Joel Emer claimed at the announcement that SMT, on the Alpha, would provide 2X the performance at a cost of less than 10% extra hardware overhead [Diefendorff, 1999].

The 21464 would have been an 8-wide superscalar, out-of-order execution, single-core processor. The move to 8-wide superscalar (which seemed inevitable in the mid 1990s) created a significant performance challenge, as few applications could effectively use that level of instruction parallelism. Thus, simultaneous multithreading became an attractive, if not essential, alternative. The 21464 contained four hardware contexts and a single unified physical register file partitioned among logical register files by register renaming. The instruction queues are fully shared. In short, this architecture shares many features of the design described in Tullsen et al. [1996], of which members of the Alpha architecture team (Joel Emer and Rebecca Stamm) were co-authors.

## 9.9 CLEARWATER NETWORKS CNP810SP

In 1998 a start-up company called XStream Logic, later Clearwater Networks, was working on the design of an SMT implementation of the MIPS IV instruction set. The processing core of Clearwater's network processor, the CNP810SP, is shown in Figure 9.4. This SMT core allows up to eight threads to execute simultaneously, utilizing a variable number of resources on a cycle by cycle

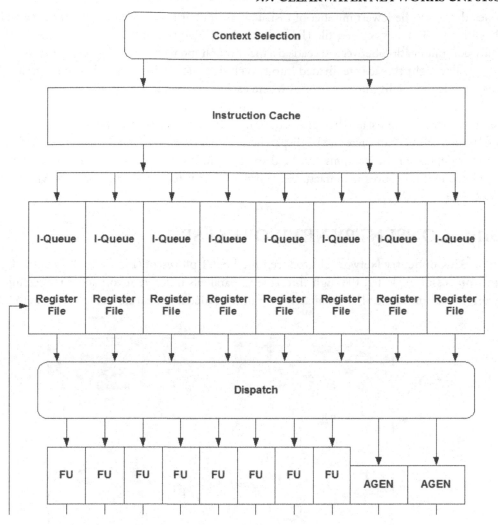

**Figure 9.4:** Clearwater CNP810SP SMT Processor consisting of 8 FU (functional unit) and 2 AGEN (address generation) units

basis. Every cycle, anywhere from zero to three instructions can be executed from each of the threads, depending on instruction dependencies and availability of resources. The maximum instructions per cycle (IPC) of the entire core is ten (eight arithmetic and two memory operations). Each cycle two threads are selected for fetch and their respective program counters (PCs) are supplied to the dual-ported instruction cache. Each port supplies eight instructions, so there is a maximum fetch bandwidth of 16 instructions. Each of the eight threads has its own instruction queue (IQ) which can hold up to 16 instructions. The two threads chosen for fetch in each cycle are the two active ones that have the fewest number of instructions in their respective IQs. Each of the eight threads has its own 31 entry register file (RF). Since there is no sharing of registers between threads, the only communication between threads occurs through memory.

The eight threads are divided into two clusters of four for ease of implementation. Thus, the dispatch logic is split into two groups where each group dispatches up to six instructions from four different threads. Eight functional units are grouped into two sets of four, each set is dedicated to a single cluster. There are also two ports to the data cache that are shared by both clusters. A maximum of 10 instructions can therefore be dispatched in each cycle. The functional units are fully bypassed so that dependent instructions can be dispatched in successive cycles. The CNP810SP was fully implemented but never manufactured, and the Clearwater design and patents where acquired by MIPS Corp. in 2001.

## 9.10   CONSENTRY NETWORKS LSP-1

In 2003, ConSentry Networks introduced the LSP-1 processor (Figure 9.5). Like the CNP810SP, this processor exploited the high thread level parallelism inherent to packet processing on a network processor. The LSP-1 is a massively multithreaded architecture that can run up to 128 threads

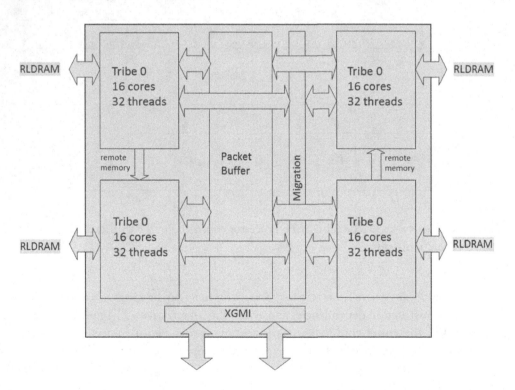

**Figure 9.5:** ConSentry LSP-1.

simultaneously, implemented as four clusters of 16 dual-threaded SMT processors. Execution resources in each core are shared between the two threads in a round robin fashion if both threads are active (not waiting on memory or blocked); if only one thread is active, it has access to all resources. Cores within a cluster (16 cores per cluster) share a dual ported ICache which can fetch up to four instructions per port. Consequently, the peak performance was 8 IPC per cluster, 32 IPC total.

## 9.11   PENTIUM 4

The Intel Pentium 4 processor [Hinton et al., 2001, Koufaty and Marr, 2003] was the first commercial, general-purpose processor to feature simultaneous multithreading, also called by the Intel marketing name hyper-threading. This processor contains two hardware contexts. It is a single-core, out-of-order scheduled, superscalar processor. It operates in two distinct modes, single-threaded and multithreaded.

As discussed in Section 6.3, there is always tension in an SMT processor between performance isolation and maximizing global performance, due to the high level of sharing. In the Pentium 4, the design navigates this issue by relying heavily on static partitioning of resources and going back and forth between the two execution modes. When in single-thread mode, that thread has full use of all resources. In multithreaded mode, each thread has access to half of the partitioned resource. This mechanism favors performance isolation, sacrificing some multithreaded performance, but without sacrificing single-thread performance. In this design (Figure 9.6), we see that the Uop queue (filled from the trace cache or the L2 cache and decoder), the instruction queues, and the reorder buffer are all statically partitioned. Instructions are scheduled out of the instruction queues independently onto the ALUs. They use a single unified physical register file, with registers identified by distinct register maps unique to each hardware context.

Intel claimed 15-27% performance gains on multithreaded and multitasking commercial workloads from turning on SMT [Koufaty and Marr, 2003]. In that paper, they indicate that the performance gains significantly outpace the hardware overhead of implementing SMT; however, they also comment that the small die area increase did not necessarily indicate a small increase in design complexity, as the first SMT implementation required the designers to think about sharing

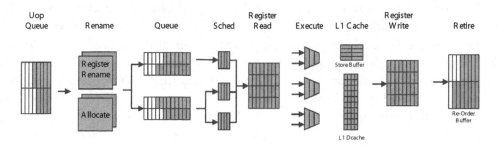

**Figure 9.6:** Part of the Pentium 4 execution pipeline.

and priority issues throughout the processor, and created new validation challenges. An independent study found 20-24% performance gain, on average, on multiprogrammed and parallel workloads for SMT on the Pentium 4 [Tuck and Tullsen, 2003].

## 9.12   SUN ULTRASPARC (NIAGARA) T1 AND T2

Sun (now Oracle) produces general purpose processor cores, but they serve a much narrower market than, for example, Intel and AMD. They are highly focused on the server market, which unlike the personal computer market, is much more defined by low per-thread ILP and high TLP. As a result, they maximize performance by maximizing many-thread throughput, and sacrifice single-thread performance. In this way, their designs are more in the spirit of the Tera MTA and other early designs than the Intel SMT designs—but only to a point, as they still maintain full bypass logic to preserve single-thread performance. The focus on throughput impacts their architecture in two ways. First, they prefer many modest cores to a few powerful cores. Second, they provide higher levels of threading per core than Intel (e.g., 4 threads on a scalar core for the Niagara T1 vs. 2 threads on a 4-wide superscalar core for Intel Nehalem). Thus, we see that even a scalar core with a short pipeline can experience this gap between hardware parallelism and software parallelism if the target workload has low ILP.

The T1 [Kongetira et al., 2005] is a chip multiprocessor with fine-grain multithreaded cores (a combination Sun dubbed chip multithreading), with elements of conjoined cores. It contains 8 cores which share an L2 cache and a single floating point unit (the conjoined core piece). Each core is an in-order, scalar core with four hardware contexts and a very simple 6 stage pipeline—fetch, thread select, decode, execute, memory, writeback. Thus, it adds one stage to accommodate multithreading—thread select comes after fetch. The thread select stage chooses an instruction from among four instruction buffers corresponding to the four hardware contexts. In this way, fetch is somewhat decoupled from the thread select, and the fetch unit's only job is to try and keep the four instruction buffers full. One of the T1 architects, James Laudon, had earlier named this combination of a scalar, pipelined machine and fine-grain multithreading *interleaved multithreading* [Laudon et al., 1994].

The Niagara T2 [Shah et al., 2007] is a chip multiprocessor of fine-grain multithreaded, conjoined cores. It has eight cores, each supporting 8 hardware contexts, for a total of 64 active threads. However, each core is really two conjoined cores (each with mostly distinct pipelines), each conjoined core supporting 4 hardware contexts. The integer pipeline is eight stages (fetch, cache, pick, decode, execute, mem, bypass, writeback). Only the fetch stage is shared between the ALU pipelines; however, the load/store pipeline stages and the floating point pipe are shared between the conjoined cores (shared by all 8 hardware contexts). Thus, integer execution and integer registers are distinct (statically partitioned to four contexts), while the instruction cache, data cache, and floating point register file are fully shared. Each of the conjoined cores is fine-grain multithreaded—one thread is selected among four each cycle in the *pick* stage in each of the two scalar pipelines. If one considers the two conjoined cores as a single core (as Sun does when they describe the T2 as an

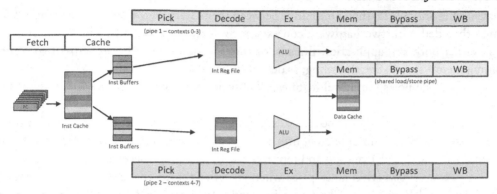

**Figure 9.7:** The conjoined integer pipeline of a Sun Niagara T2 core. The fetch unit, load/store pipe, and floating point pipe are shared, but the integer execution pipelines are distinct and each fine-grain multithreaded.

eight-core processor), then it is marginally simultaneous multithreaded in the sense that two threads can have instructions in the execute (for example) stage at the same time. However, it lacks the full flexibility of SMT—no thread can use both execution units at once. So it is more accurate to consider it two conjoined fine-grain pipelines.

## 9.13   SUN MAJC AND ROCK

While not a commercial success, the Sun MAJC [Tremblay et al., 2000] architecture included some unique and aggressive features. It is a processor designed exclusively for execution of Java code. It is a conjoined core, (4 operation wide) VLIW processor with support for (4-context) coarse-grain multithreading. It is a conjoined core, because each core features a distinct instruction cache, VLIW pipeline, and execution units but a single shared data cache, connection to the interconnect, etc.

MAJC features support for speculative multithreading, allowing the parallelization of loops that are not provably free of loop-carried dependencies. Parallel threads execute distinct iterations of a loop which are assumed to be independent, and later iterations are squashed if a violation is detected. They introduce *virtual channels* to handle register dependencies, and manage memory-based dependencies in software. Virtual channels allow the compiler to explicitly pass register values between threads, somewhat reminiscent of Multiscalar [Sohi et al., 1998]. Software-based detection of memory ordering violations can be quite expensive, but they exploit a feature in Java that keeps private (stack) data distinct from shared (heap) data, allowing them to reduce that overhead.

Sun's ROCK processor [Chaudhry et al., 2009a] was even less of a commercial success (canceled before launch), but also exceeded MAJC (and other contemporary processors) in threading-based innovation. The processor described by Sun in the research literature was to have four CMP cores, each CMP core composed of four conjoined cores. The four core "cluster" is conjoined because the four pipelines share a single instruction fetch unit and instruction cache as well as two data caches and two floating point units between them. Each of the 16 in-order cores is fine-grained multithreaded with two hardware contexts, for a total of 32 threads. The two threads on a core can either both run application threads, or run one application thread across two contexts using simultaneous speculative threading, as described below.

Rock supports four novel features, all of which heavily leverage the architects' choice to support fast, frequent checkpointing. A checkpoint captures microarchitectural state (program counter, registers, etc.) at a particular point in execution, so that the thread can continue in some kind of speculative manner, and either confirm the speculative execution (and dump the checkpoint) or dump the new speculative state and return to the checkpoint if necessary.

*Execute Ahead* (EA) takes place when an active thread encounters a long-latency operation (cache miss, TLB miss, even a floating point divide). In this mode, a checkpoint is taken and execution continues. Instructions dependent on the stalled instruction are placed in the deferred instruction queue (DQ) for later execution, but independent instructions complete and speculatively retire. When the stalled instruction completes, the processor stops EA execution and begins completing the instructions in the DQ until it catches up. This optimization has two primary advantages. (1) It keeps the execution units busy when a conventional processor would be blocked, giving the in-order processor out-of-order capabilities in the presence of long-latency operations, and (2) it can significantly increase memory-level parallelism, initiating future cache misses before the earlier misses are resolved. If the EA mode fails (because the store buffer or DQ overflow), the thread goes into *hardware scout* mode, where independent instructions execute but do not retire (similar to runahead execution [Mutlu et al., 2003]), and execution restarts at the checkpoint when the stalled instruction completes. In this mode, execution only benefits from the latter effect (memory-level parallelism). Both of these optimizations can operate without using a different hardware context than the application thread (due to the checkpoint support).

If we are willing to use both hardware contexts, Rock can exploit *simultaneous speculative threading* (SST) [Chaudhry et al., 2009b]. With SST, the leading thread no longer needs to stop when the stalled instruction completes; instead, both threads continue to move forward, with the leading thread continuing to possibly encounter misses, create new checkpoints, and defer more instructions. Thus, the two threads operate in parallel.

Finally, Rock is the first processor to support *transactional memory* [Herlihy and Moss, 1993] in hardware. Code in a transaction either completes atomically, or fails and must restart. Transactional memory allows the programmer or compiler to speculatively parallelize code that cannot be guaranteed to be free of dependencies, depending on the hardware to detect violations when they occur. This increases the availability of threads and parallelism in code.

Marc Tremblay served as chief architect for both MAJC and Rock, as well as other more successful multithreaded (and single-threaded) architectures for Sun.

# 9.14   IBM POWER

The IBM Power5 [Kalla et al., 2004] is a dual-core chip multiprocessor of simultaneous multi-threaded cores. Each core supports two hardware contexts. The pipeline of a single core is shown in Figure 9.8. Each cycle, the core fetches from one thread (up to eight instructions) into one of the two instruction buffers, each dedicated to a different hardware context. Instructions are dispatched (up to 5 instructions per cycle) from one of the two instruction buffers, and assembled into a group—a group forms a set of instructions that will eventually be committed together. This group is renamed and placed in the appropriate instruction queues. Up until this point, there is no mixing of instructions in a pipeline stage – SMT mixing of instructions only happens as instructions are issued to the functional units out of the queues.

One goal of the Power5 was to maximize compatibility with the Power4, even to the extent that pipeline-based software optimizations would still apply on the Power5. This makes the Power5 one of the most interesting simultaneous multithreading architectures, because we can make the most direct before-and-after comparisons. This is because very little changed between the Power4 and Power5 architectures except the addition of multithreading. For example, we see that the architects were able to implement simultaneous multithreading without adding a single pipeline stage. The new resources added were an extra program counter, an extra rename mapper, and duplicated completion logic. They increased the number of renaming registers (from 80 to 120 integer registers and from 70 to 120 floating point registers), and increased the associativity of the L1 instruction cache (from direct-mapped to 2-way) and data cache (from 2-way to 4-way).

The Power5 features explicit support for thread priority. Although the core only dispatches from one thread per cycle, the thread priority determines how often each thread gets a dispatch cycle. This mechanism can be used to enforce, for example, Linux execution priorities or to give priority to a real-time application over a latency-insensitive application [Boneti et al., 2008]. Perhaps most useful, it can be used to minimize resources used by a thread spinning in the idle loop or waiting for synchronization.

However, the hardware also has some control over the relative dispatch rates. This allows the processor to implement mechanisms that prevent one thread from monopolizing pipeline resources when it is stalled. Depending on the source of the stall that is causing a thread to clog resources, the hardware can impose three different levels of throttling. The hardware can automatically reduce the priority of a thread that is using too many resources, it can stop fetch of a thread that incurs too many cache misses, and can even flush instructions when a thread is stalled waiting for synchronization (which can incur an unbounded latency). The first mechanism is similar to ICOUNT [Tullsen et al., 1996], the second and third are similar to STALL and FLUSH [Tullsen and Brown, 2001], respectively.

Like the Pentium 4, the Power5 also has the ability to transition to a special single-thread mode. The Power5 uses less static partitioning than the Pentium 4, so most resources become fully available to a lone active thread, even in SMT mode. The exception, and the primary advantage of single-thread mode, is that physical registers that would have to hold the architectural register state

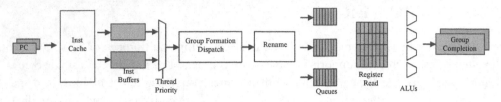

**Figure 9.8:** The IBM Power5 execution pipeline.

of even an idle thread can be freed and made available for renaming in single-thread mode. Thus, an extra 32 or so registers become available in this mode.

The IBM Power6 [Le et al., 2007], like the Power5, is a dual-core CMP of SMT processors, with each core supporting two threads. The biggest difference between the two is that Power6 pipelines are in-order, allowing them to eliminate register renaming and other complex functions, and thereby enable more aggressive clock rates. This is a trade-off (higher clock-rate at the expense of lower per-thread ILP) that is considerably more attractive with SMT to supplement the ILP available in a single thread. SMT issue is challenging at high clock rates—the architecture thus again exploits the instruction grouping we saw in the Power5. It puts fetched instructions into groups (without inter-instruction dependencies and no internal branches) that can safely be dispatched together, up to 5 instructions. Only if the next group from both threads do not exceed the total execution resources (7 instruction issue) can both issue in the same cycle.

The IBM Power7 [Sinharoy et al., 2011], attempting to maximize performance/energy, backed off of the Power6 approach, trading higher ILP for a lower clock rate. High ILP is achieved by re-introducing out-of-order execution, and increasing threads per core to 4 on a wide (6-issue) superscalar. Many shared resources are partitioned, thus a core explicitly executes in one of three modes—single-thread (ST), dual thread (SMT2), or four thread (SMT4). A core, in fact, looks more like two pipelines, each with its own register file. In ST and SMT2 mode, the register files are kept identical, allowing threads to freely dispatch instructions to either pipe. In SMT4, each register file and pipeline is distinct, servicing two threads. Thus, this architecture can go back and forth from what we would call a *clustered architecture* (multiple pipelines and threads using them both) in ST and SMT2 mode to a conjoined core architecture (multiple pipes, threads constrained to one or the other) in SMT4 mode.

## 9.15   AMD BULLDOZER

Although AMD has yet to produce its first truly multithreaded core, the Bulldozer architecture [Butler et al., 2011] is a (2 to 4 core) CMP of conjoined cores. Each core is a 4-wide superscalar out-of-order pipeline, conjoined but with aspects of fine-grain multithreading and SMT. Each core supports two threads, but in a conjoined manner with two integer pipelines for the exclusive use of each thread. However, the two pipelines share the front end, including instruction cache, branch prediction, and decode. They also share the floating point pipeline, which actually employs simultaneous multithreading instruction issue. The pipelines each have their own data cache. This architecture is probably the closest commercial architecture to the original conjoined core proposal [Kumar et al., 2004].

## 9.16    INTEL NEHALEM

The Core i7 instantiation of the Nehalem architecture [Dixon et al., 2010] is a four-core CMP of SMT cores. Each core contains two hardware contexts and a 4-wide superscalar out-of-order pipeline. Nehalem-EX architectures extend up to eight cores. The architectural description distinguishes between resources that are fully replicated and private to each core (register state, the return stack, and the large-page ITLB), resources that are partitioned when two threads are active (load buffer, store buffer, reorder buffer, small-page ITLB), completely shared (reservation stations, caches, DTLB), and completely unaware of threading (execution units). While many resources are still partitioned in this architecture, it is still much more flexible than the Pentium 4, because the reservation stations are fully shared, and the designers allow most of the partitioned resources to become fully available to a single thread, even in SMT mode, when the other thread is inactive.

## 9.17    SUMMARY

Table 9.1 summarizes processor implementations discussed in this Chapter.

**Table 9.1:** Multithreaded Processors

| Processor | Technologies | Cores | Threads/Core | Notes |
|---|---|---|---|---|
| NBS DYSEAC | CGMT | 1 | 2 | First CGMT Processor |
| MIT Lincoln Lab TX-2 | CGMT | 1 | 33 | |
| CDC 6600 | FGMT | 1 | 10 | First FGMT Processor |
| Denelcor HEP | CMP, FGMT | 16 | 128 | |
| Horizon | FGMT | 1 | 128 | |
| Tera MTA | CMP, FGMT | 4 | 128 | |
| Delco TIO | FGMT | 2 | 32 | First RT FGMT Processor |
| MIT Sparcle | CGMT | 1 | 4 | modified SPARC |
| Alpha 21464 | SMT | 1 | 4 | |
| Clearwater CNP810SP | SMT | 1 | 8 | |
| Consentry LSP-1 | CMP, FGMT | 64 | 2 | |
| Intel Pentium 4 | SMT | 1 | 2 | First GP SMT |
| Sun Niagara T1 | CMP, FGMT | 8 | 4 | |
| Sun Niagara T2 | CMP, FGMT | 8 | 8 | |
| Sun MAJC 5200 | CMP, CGMT | 2 | 4 | speculative multithreading |
| Sun Rock | CMP, FGMT, conjoined | 4 | 8 | |
| IBM Power5 | CMP, SMT | 2 | 2 | out-of-order |
| IBM Power6 | CMP, SMT | 2 | 2 | in-order |
| IBM Power7 | CMP, SMT | 8 | 4 | out-of-order |
| AMD Bulldozer | CMP, conjoined | 4 | 2 | |
| Intel Nehalem i7 | CMP, SMT | 4 | 2 | |

# Bibliography

Tor M. Aamodt, Paul Chow, Per Hammarlund, Hong Wang, and John P. Shen. Hardware support for prescient instruction prefetch. In *Proceedings of the International Symposium on High-Performance Computer Architecture*, pages 84–95, 2004. 55

Onur Aciicmez and Jean-Pierre Seifert. Cheap hardware parallelism implies cheap security. In *Proceedings of the Workshop on Fault Diagnosis and Tolerance in Cryptography*, pages 80–91, 2007. 50

A. Agarwal, J. Kubiatowicz, D. Kranz, B-H. Lim, D. Yeung, G. D'Souza, and M. Parkin. Sparcle: An evolutionary processor design for large-scale multiprocessors. *IEEE Micro*, June 1993. 10, 18, 72

Anant Agarwal, Ricardo Bianchini, David Chaiken, Kirk L. Johnson, David Kranz, John Kubiatowicz, Beng-Hong Lim, Kenneth Mackenzie, and Donald Yeung. The MIT Alewife machine: architecture and performance. In *Proceedings of the International Symposium on Computer Architecture*, pages 2–13, 1995. 9, 18, 50

Haitham Akkary and Michael A. Driscoll. A dynamic multithreading processor. In *Proceedings of the International Symposium on Microarchitecture*, pages 226–236, 1998. 57

Robert Alverson, David Callahan, Daniel Cummings, Brian Koblenz, Allan Porterfield, and Burton Smith. The Tera computer system. In *Proceedings of the International Conference on Supercomputing*, pages 1–6, 1990. 12, 25, 50, 71

AMD. AMD Phenom II processors. URL http://www.amd.com/us/products/desktop/processors/phenom-ii/Pages/phenom-ii.aspx. 8

Thomas E. Anderson. The performance of spin lock alternatives for shared-money multiprocessors. *IEEE Transactions on Parallel and Distributed Systems*, 1:6–16, 1990. 50

Robert W. Bemer. How to consider a computer. *Automatic Control*, pages 66–69, March 1957. 17

Carlos Boneti, Francisco J. Cazorla, Roberto Gioiosa, Alper Buyuktosunoglu, Chen-Yong Cher, and Mateo Valero. Software-controlled priority characterization of POWER5 processor. In *Proceedings of the International Symposium on Computer Architecture*, pages 415–426, 2008. 79

Eric Borch, Srilatha Manne, Joel Emer, and Eric Tune. Loose loops sink chips. In *Proceedings of the International Symposium on High-Performance Computer Architecture*, 2002. 47

Jeffery A. Brown and Dean M. Tullsen. The shared-thread multiprocessor. In *Proceedings of the International Conference on Supercomputing*, pages 73–82, 2008. 15

Michael Butler, Leslie Barnes, Debjit Das Sarma, and Bob Gelinas. Bulldozer: An approach to multithreaded compute performance. *IEEE Micro*, 31(2):6–15, March 2011. 8, 80

J. Adam Butts and Gurindar S. Sohi. Use-based register caching with decoupled indexing. In *Proceedings of the International Symposium on Computer Architecture*, 2004. 47

Francisco Cazorla, Alex Ramirez, Mateo Valero, and Enrique Fernandez. Dynamically controlled resource allocation in SMT processors. In *Proceedings of the International Symposium on Microarchitecture*, pages 171–182, 2004a. 46

Francisco J. Cazorla, Alex Ramirez, Mateo Valero, and Enrique Fernandez. DCache warn: an I-fetch policy to increase SMT efficiency. In *Proceedings of the International Parallel and Distributed Processing Symposium*, pages 74–83, 2004b.

Francisco J. Cazorla, Alex Pajuelo, Oliverio J. Santana, Enrique Fernandez, and Mateo Valero. On the problem of evaluating the performance of multiprogrammed workloads. *IEEE Transactions on Computers*, 59(12):1722 –1728, dec. 2010. 65

Robert S. Chappell, Jared Stark, Sangwook P. Kim, Steven K. Reinhardt, and Yale N. Patt. Simultaneous subordinate microthreading (SSMT). In *Proceedings of the International Symposium on Computer Architecture*, pages 186–195, 1999. 54, 55

Shailender Chaudhry, Robert Cypher, Magnus Ekman, Martin Karlsson, Anders Landin, Sherman Yip, Hakan Zeffer, and Marc Tremblay. Rock: A high-performance SPARC CMT processor. *IEEE Micro*, 29:6–16, 2009a. 77

Shailender Chaudhry, Robert Cypher, Magnus Ekman, Martin Karlsson, Anders Landin, Sherman Yip, Håkan Zeffer, and Marc Tremblay. Simultaneous speculative threading: a novel pipeline architecture implemented in Sun's Rock processor. In *Proceedings of the International Symposium on Computer Architecture*, pages 484–495, 2009b. 78

Bumyong Choi, Leo Porter, and Dean M. Tullsen. Accurate branch prediction for short threads. In *Proceedings of the Symposium on Architectural Support for Programming Languages and Operating Systems*, pages 125–134, 2008. 44

Seungryul Choi and Donald Yeung. Learning-based SMT processor resource distribution via hill-climbing. In *Proceedings of the International Symposium on Computer Architecture*, pages 239–251, 2006. 46

Jamison D. Collins, Dean M. Tullsen, Hong Wang, and John P. Shen. Dynamic speculative precomputation. In *Proceedings of the International Symposium on Microarchitecture*, pages 306–317, 2001a. 54

Jamison D. Collins, Hong Wang, Dean M. Tullsen, Christopher Hughes, Yong-Fong Lee, Dan Lavery, and John P. Shen. Speculative precomputation: long-range prefetching of delinquent loads. In *Proceedings of the International Symposium on Computer Architecture*, pages 14–25, 2001b. 54, 55

Adrián Cristal, Oliverio J. Santana, Mateo Valero, and José F. Martínez. Toward kilo-instruction processors. *ACM Transactions on Architecture and Code Optimimization*, 1(4):389–417, December 2004. 44

José-Lorenzo Cruz, Antonio González, Mateo Valero, and Nigel P. Topham. Multiple-banked register file architectures. In *Proceedings of the International Symposium on Computer Architecture*, pages 316–325, 2000. 47

John Demme, Robert Martin, Adam Waksman, and Simha Sethumadhavan. Side-channel vulnerability factor: a metric for measuring information leakage. In *Proceedings of the International Symposium on Computer Architecture*, pages 106–117, 2012. 51

Matthew Devuyst, Rakesh Kumar, and Dean M. Tullsen. Exploiting unbalanced thread scheduling for energy and performance on a CMP of SMT processors. In *Proceedings of the International Parallel and Distributed Processing Symposium*, 2006. 48

Keith Diefendorff. Compaq chooses SMT for Alpha. In *Microprocessor Report, Vol 13, No. 16*, December 1999. 13, 33, 72

Martin Dixon, Per Hammarlund, Stephan Jourdan, and Ronak Singhai. The next generation Intel core microarchitecture. *Intel Technology Journal*, 14(3), 2010. 13, 34, 81

Gautham K. Dorai and Donald Yeung. Transparent threads: Resource sharing in SMT processors for high single-thread performance. In *Proceedings of the International Conference on Parallel Architectures and Compilation Techniques*, 2002. 48

Ali El-Moursy and David H. Albonesi. Front-end policies for improved issue efficiency in SMT processors. In *Proceedings of the International Symposium on High-Performance Computer Architecture*, pages 31–40, 2003. 46

Joel Emer. Simultaneous multithreading: Multiplying Alpha performance. In *Proceedings of Microprocessor Forum*, 1999. 33, 72

Oguz Ergin, Deniz Balkan, Kanad Ghose, and Dmitry Ponomarev. Register packing: Exploiting narrow-width operands for reducing register file pressure. In *Proceedings of the International Symposium on Microarchitecture*, pages 304–315, 2004. 47

Stijn Eyerman and Lieven Eeckhout. A memory-level parallelism aware fetch policy for SMT processors. In *Proceedings of the International Symposium on High-Performance Computer Architecture*, pages 240–249, 2007. 46

Stijn Eyerman and Lieven Eeckhout. System-level performance metrics for multiprogram workloads. *IEEE Micro*, 28(3):42 –53, May 2008. 64

Stijn Eyerman and Lieven Eeckhout. Probabilistic job symbiosis modeling for SMT processor scheduling. In *Proceedings of the Symposium on Architectural Support for Programming Languages and Operating Systems*, pages 91–102, 2010. 48

Keith I. Farkas and Norm P. Jouppi. Complexity/performance tradeoffs with non-blocking loads. In *Proceedings of the International Symposium on Computer Architecture*, pages 211–222, 1994. 21

Alexandra Fedorova, Margo Seltzer, Christoper Small, and Daniel Nussbaum. Performance of multithreaded chip multiprocessors and implications for operating system design. In *Proceedings of the USENIX Annual Technical Conference*, pages 26–26, 2005. 48

Michael J. Flynn. Some computer organizations and their effectiveness. *IEEE Transactions on Computers*, 21(9):948–960, September 1972. 16

James W. Forgie. The lincoln TX-2 input-output system. In *Proceedings of the Western Joint Computer Conference*, pages 156–160, 1957. 17

Ron Gabor, Shlomo Weiss, and Avi Mendelson. Fairness enforcement in switch on event multithreading. *ACM Transactions on Architecture and Code Optimization*, 4(3), September 2007. 65

Simcha Gochman, Avi Mendelson, Alon Naveh, and Efraim Rotem. Introduction to Intel Core Duo processor architecture. *Intel Technology Journal*, May 2006. 8

Manu Gulati and Nader Bagherzadeh. Performance study of a multithreaded superscalar microprocessor. In *Proceedings of the International Symposium on High-Performance Computer Architecture*, pages 291–301, 1996. 12, 33, 39

Maurice Herlihy and J. Eliot B. Moss. Transactional memory: architectural support for lock-free data structures. In *Proceedings of the International Symposium on Computer Architecture*, pages 289–300, 1993. 78

Sebastien Hily and Andre Seznec. Branch prediction and simultaneous multithreading. In *Proceedings of the International Conference on Parallel Architectures and Compilation Techniques*, 1996. 44

Sebastien Hily and Andre Seznec. Standard memory hierarchy does not fit simultaneous multithreading. In *Workshop on Multithreaded Execution Architecture and Compilation*, January 1998. 42

Sebastien Hily and Andre Seznec. Out-of-order execution may not be cost-effective on processors featuring simultaneous multithreading. In *Proceedings of the International Symposium on High-Performance Computer Architecture*, 1999. 39

Glenn Hinton, Dave Sager, Mike Upton, Darrell Boggs, Doug Carmean, Alan Kyker, and Patrice Roussel. The microarchitecture of the Pentium 4 processor. *Intel Technology Journal*, Feb 2001. 13, 34, 75

Hiroaki Hirata, Kozo Kimura, Satoshi Nagamine, Yoshiyuki Mochizuki, Akio Nishimura, Yoshimori Nakase, and Teiji Nishizawa. An elementary processor architecture with simultaneous instruction issuing from multiple threads. In *Proceedings of the International Symposium on Computer Architecture*, pages 136–145, May 1992. 12, 33, 39

Timothy M. Jones, Michael F. P. O'Boyle, Jaume Abella, Antonio Gonzalez, and Oğuz Ergin. Compiler directed early register release. In *Proceedings of the International Conference on Parallel Architectures and Compilation Techniques*, pages 110–122, 2005. 47

Stephen Jourdan, Ronny Ronen, Michael Bekerman, Bishara Shomar, and Adi Yoaz. A novel renaming scheme to exploit value temporal locality through physical register reuse and unification. In *Proceedings of the International Symposium on Microarchitecture*, pages 216–225, 1998. 47

Ron Kalla, Balaram Sinharoy, and Joel M. Tendler. IBM Power5 chip: A dual-core multithreaded processor. *IEEE Micro*, 24(2):40–47, March 2004. 13, 34, 37, 79

W.J. Kaminsky and Edward S. Davidson. Special feature: Developing a multiple-instructon-stream single-chip processor. *Computer*, 12:66–76, 1979. 26

Gerry Kane and Joe Heinrich. *MIPS RISC Architecture*. Prentice Hall, 1992. 18

Stephem W. Keckler and William J. Dally. Processor coupling: integrating compile time and runtime scheduling for parallelism. In *Proceedings of the International Symposium on Computer Architecture*, pages 202–213, 1992. 12, 33, 39

Stephen W. Keckler, William J. Dally, Daniel Maskit, Nicholas P. Carter, Andrew Chang, and Whay S. Lee. Exploiting fine-grain thread level parallelism on the MIT multi-ALU processor. In *Proceedings of the International Symposium on Computer Architecture*, pages 306–317, 1998. 50

J. Kihm, A. Settle, A. Janiszewski, and D.A. Connors. Understanding the impact of inter-thread cache interference on ILP in modern SMT processors. *Journal of Instruction Level Parallelism*, 7 (2), 2005. 48

Dongkeun Kim and Donald Yeung. Design and evaluation of compiler algorithms for pre-execution. In *Proceedings of the Symposium on Architectural Support for Programming Languages and Operating Systems*, pages 159–170, 2002. 54

Hyesoon Kim, Richard Vuduc, Sara Baghsorkhi, Jee Choi, and Wen-mei Hwu. *Performance Analysis and Tuning for General Purpose Graphics Processing Units (GPGPU)*, volume 7(2) of *Synthesis Lectures on Computer Architecture*. Morgan & Claypool Publishers, 2012. 16

Seongbeom Kim, Dhruba Chandra, and Yan Solihin. Fair cache sharing and partitioning in a chip multiprocessor architecture. In *Proceedings of the International Conference on Parallel Architectures and Compilation Techniques*, pages 111–122, 2004. 65

Poonacha Kongetira, Kathirgamar Aingaran, and Kunle Olukotun. Niagara: A 32-way multi-threaded SPARC processor. *IEEE MICRO*, March 2005. 12, 26, 76

D. Koufaty and D.T. Marr. Hyperthreading technology in the netburst microarchitecture. *IEEE Micro*, pages 56–65, April 2003. 33, 34, 42, 75

David Kroft. Lockup-free instruction fetch/prefetch cache organization. In *Proceedings of the International Symposium on Computer Architecture*, pages 81–87, 1981. 21

J. T. Kuehn and B. J. Smith. The Horizon supercomputing system: architecture and software. In *Proceedings of the International Conference on Supercomputing*, pages 28–34, 1988. 70

Rakesh Kumar and Dean M. Tullsen. Compiling for instruction cache performance on a multi-threaded architecture. In *Proceedings of the International Symposium on Microarchitecture*, pages 419–429, 2002. 49

Rakesh Kumar, Norman P. Jouppi, and Dean M. Tullsen. Conjoined-core chip multiprocessing. In *Proceedings of the International Symposium on Microarchitecture*, pages 195–206, 2004. 8, 81

Butler W. Lampson and Kenneth A. Pier. A processor for a high-performance personal computer. In *Proceedings of the International Symposium on Computer Architecture*, pages 146–160, 1980. 9, 17

J. Laudon, A. Gupta, and M. Horowitz. Interleaving: A multithreading technique targeting multipro-cessors and workstations. In *Proceedings of the Symposium on Architectural Support for Programming Languages and Operating Systems*, pages 308–318, October 1994. 76

H. Q. Le, W. J. Starke, J. S. Fields, F. P. O'Connell, D. Q. Nguyen, B. J. Ronchetti, W. M. Sauer, E. M. Schwarz, and M. T. Vaden. IBM POWER6 microarchitecture. *IBM Journal of Research and Development*, 51(6), November 2007. 80

Alan L. Leiner. System specifications for the DYSEAC. *Journal of the ACM*, 1(2):57–81, April 1954. 17, 67

Yingmin Li, D. Brooks, Zhigang Hu, K. Skadron, and P. Bose. Understanding the energy efficiency of simultaneous multithreading. In *International Symposium on Low Power Electronics and Design*, pages 44 –49, August 2004. 58

Mikko H. Lipasti, Brian R. Mestan, and Erika Gunadi. Physical register inlining. In *Proceedings of the International Symposium on Computer Architecture*, 2004. 47

Jack L. Lo, Susan J. Eggers, Henry M. Levy, Sujay S. Parekh, and Dean M. Tullsen. Tuning compiler optimizations for simultaneous multithreading. In *Proceedings of the International Symposium on Microarchitecture*, pages 114–124, 1997. 49

Jack L. Lo, Sujay S. Parekh, Susan J. Eggers, Henry M. Levy, and Dean M. Tullisen. Software-directed register deallocation for simultaneous multithreaded processors. *IEEE Transactions on Parallel and Distributed Systems*, 10(9):922–933, September 1999. 47

S. López, S. Dropsho, D. Albonesi, O. Garnica, and J. Lanchares. Dynamic capacity-speed tradeoffs in SMT processor caches. *High Performance Embedded Architectures and Compilers*, pages 136–150, 2007. 42

Chi-Keung Luk. Tolerating memory latency through software-controlled pre-execution in simultaneous multithreading processors. In *Proceedings of the International Symposium on Computer Architecture*, pages 40–51, 2001. 54, 55

Kun Luo, J. Gummaraju, and M. Franklin. Balancing thoughput and fairness in SMT processors. In *International Symposium on Performance Analysis of Systems and Software*, pages 164–171, 2001. 64

Carlos Madriles, Carlos García-Quiñones, Jesús Sánchez, Pedro Marcuello, Antonio González, Dean M. Tullsen, Hong Wang, and John P. Shen. Mitosis: A speculative multithreaded processor based on precomputation slices. *IEEE Transactions on Parallel and Distributed Systems*, 19(7): 914–925, July 2008. 55

Pedro Marcuello and Antonio González. Exploiting speculative thread-level parallelism on a SMT processor. In *Proceedings of the International Conference on High-Performance Computing and Networking*, pages 754–763, 1999. 57

Pedro Marcuello, Antonio González, and Jordi Tubella. Speculative multithreaded processors. In *Proceedings of the International Conference on Supercomputing*, pages 77–84, 1998. 56

Deborah T. Marr. Microarchitectue choices and tradeoffs for maximizing processing efficiency. PhD Thesis, University of Michigan, 2008. 59

José F. Martínez, Jose Renau, Michael C. Huang, Milos Prvulovic, and Josep Torrellas. Cherry: checkpointed early resource recycling in out-of-order microprocessors. In *Proceedings of the International Symposium on Microarchitecture*, pages 3–14, 2002. 44

Teresa Monreal, Víctor Viñals, Antonio González, and Mateo Valero. Hardware schemes for early register release. In *Proceedings of the International Conference on Parallel Processing*, 2002. 47

Onur Mutlu, Jared Stark, Chris Wilkerson, and Yale N. Patt. Runahead execution: An alternative to very large instruction windows for out-of-order processors. In *Proceedings of the International Symposium on High-Performance Computer Architecture*, 2003. 78

Mario Nemirovsky and Matthew Sale. Microprogrammed timer processor, May 1992a. US Patent 5,117,387. 25, 70

Mario Nemirovsky and Matthew Sale. Microprogrammed timer processor, May 1992b. US Patent 5,115,513. 70

Mario Nemirovsky and Wayne Yamamoto. Quantitative study of data caches on a multistreamed architecture. In *Workshop on Multithreaded Execution Architecture and Compilation*, January 1998. 42

Dimitrios Nikolopoulos. Code and data transformations for improving shared cache performance on SMT processors. In *High Performance Computing*, volume 2858 of *Lecture Notes in Computer Science*, pages 54–69. Springer, 2003. 49

David W. Oehmke, Nathan L. Binkert, Trevor Mudge, and Steven K. Reinhardt. How to fake 1000 registers. In *Proceedings of the International Symposium on Microarchitecture*, pages 7–18, 2005. 47

Kunle Olukotun, Basem A. Nayfeh, Lance Hammond, Ken Wilson, and Kunyung Chang. The case for a single-chip multiprocessor. In *Proceedings of the Symposium on Architectural Support for Programming Languages and Operating Systems*, pages 2–11, 1996. 7

Kunle Olukotun, Lance Hammond, and James Laudon. Chip multiprocessor architecture: Techniques to improve throughput and latency. 2(1)(1), 2007. 8

Venkatesan Packirisamy, Yangchun Luo, Wei-Lung Hung, Antonia Zhai, Pen-Chung Yew, and Tin-Fook Ngai. Efficiency of thread-level speculation in SMT and CMP architectures - performance, power and thermal perspective. In *International Conference on Computer Design*, October 2008. 57

Il Park, Babak Falsafi, and T. N. Vijaykumar. Implicitly-multithreaded processors. In *Proceedings of the International Symposium on Computer Architecture*, pages 39–51, 2003. 57

David A. Patterson and Carlo H. Sequin. RISC I: A Reduced Instruction Set VLSI Computer. In *Proceedings of the International Symposium on Computer Architecture*, pages 443–457, 1981. 10

Michael D. Powell, Mohamed Gomaa, and T. N. Vijaykumar. Heat-and-run: leveraging SMT and CMP to manage power density through the operating system. In *Proceedings of the Symposium on Architectural Support for Programming Languages and Operating Systems*, pages 260–270, 2004. 48

Matt Ramsay, Chris Feucht, and Mikko H. Lipasti. Exploring efficient SMT branch predictor design. 2003. 44

Joshua Redstone, Susan Eggers, and Henry Levy. Mini-threads: Increasing TLP on small-scale SMT processors. In *Proceedings of the International Symposium on High-Performance Computer Architecture*, 2003. 47

Joshua A. Redstone, Susan J. Eggers, and Henry M. Levy. An analysis of operating system behavior on a simultaneous multithreaded architecture. In *In Proceedings of the 9th International Conference on Architectural Support for Programming Languages and Operating Systems*, pages 245–256, 2000. 47

Steven K. Reinhardt and Shubhendu S. Mukherjee. Transient fault detection via simultaneous multithreading. In *Proceedings of the International Symposium on Computer Architecture*, pages 25–36, 2000. 56

Eric Rotenberg. AR-SMT: A microarchitectural approach to fault tolerance in microprocessors. In *Proceedings of the International Symposium on Fault-Tolerant Computing*, pages 84–91, 1999. 56

Larry Rudolph and Zary Segall. Dynamic decentralized cache schemes for MIMD parallel processors. In *Proceedings of the International Symposium on Computer Architecture*, pages 340–347, 1984. 50

Subhradyuti Sarkar and Dean M. Tullsen. Compiler techniques for reducing data cache miss rate on a multithreaded architecture. In *Proceedings of the International Conference on High Performance Embedded Architectures and Compilers*, pages 353–368, 2008. 49

Robert Schöne, Daniel Hackenberg, and Daniel Molka. Simultaneous multithreading on x86_64 systems: an energy efficiency evaluation. In *Workshop on Power-Aware Computing and Systems (HotPower)*, 2011. 59

John S. Seng, Dean M. Tullsen, and George Z.N. Cai. Power-sensitive multithreaded architecture. In *International Conference on Computer Design*, 2000. 40, 58

Mauricio Serrano, Roger Wood, and Mario Nemirovsky. A study on multistreamed superscalar processors. Technical report, University of California Santa Barbara, May 1993. 33

Mauricio Jose Serrano. Performance tradeoffs in multistreamed superscalar architectures. PhD Thesis, University of California, Santa Barbara, 1994. 44

A. Settle, D. Connors, E. Gibert, and A. González. A dynamically reconfigurable cache for multi-threaded processors. *Journal of Embedded Computing*, 2(2):221–233, 2006. 43

M. Shah, J. Barren, J. Brooks, R. Golla, G. Grohoski, N. Gura, R. Hetherington, P. Jordan, M. Luttrell, C. Olson, B. Sana, D. Sheahan, L. Spracklen, and A. Wynn. UltraSPARC T2: A highly-treaded, power-efficient, SPARC SOC. In *IEEE Asian Solid-State Circuits Conference*, November 2007. 8, 12, 76

Harsh Sharangpani and Ken Arora. Itanium processor microarchitecture. *IEEE Micro*, 20(5):24–43, September 2000. 39

T. Sherwood, E. Perelman, G. Hammerley, and B. Calder. Automatically characterizing large-scale program behavior. In *International Conference on Architectural Support for Programming Languages and Operating Systems*, October 2002. 64

B. Sinharoy, R. Kalla, W. J. Starke, H. Q. Le, R. Cargnoni, J. A. Van Norstrand, B. J. Ronchetti, J. Stuecheli, J. Leenstra, G. L. Guthrie, D. Q. Nguyen, B. Blaner, C. F. Marino, E. Retter, and P. Williams. IBM POWER7 multicore server processor. *IBM Journal of Research and Development*, 55(3):191–219, May 2011. 80

Burton J. Smith. Architecture and applications of the HEP mulitprocessor computer system. In *Proceedings of the International Society for Optical Engineering*, 1982. 25, 50, 68

J. E. Smith. Characterizing computer performance with a single number. *Communications of the ACM*, 31(10):1202–1206, October 1988. 64

Allan Snavely and Dean M. Tullsen. Symbiotic jobscheduling for a simultaneous multithreading architecture. In *Proceedings of the Symposium on Architectural Support for Programming Languages and Operating Systems*, November 2000. 47, 63

Allan Snavely, Dean M. Tullsen, and Geoff Voelker. Symbiotic jobscheduling with priorities for a simultaneous multithreading processor. In *Proceedings of the ACM SIGMETRICS Conference on Measurement and Modeling of Computer Systems*, pages 66–76, 2002. 48

Gurindar S. Sohi, Scott E. Breach, and T. N. Vijaykumar. Multiscalar processors. In *25 Years ISCA: Retrospectives and Reprints*, pages 521–532, 1998. 56, 77

Lawrence Spracklen and Santosh G. Abraham. Chip multithreading: Opportunities and challenges. In *Proceedings of the International Symposium on High-Performance Computer Architecture*, pages 248–252, 2005. 13

G. Edward Suh, Srinivas Devadas, and Larry Rudolph. A new memory monitoring scheme for memory-aware scheduling and partitioning. In *Proceedings of the International Symposium on High-Performance Computer Architecture*, 2002. 43

Karthik Sundaramoorthy, Zach Purser, and Eric Rotenburg. Slipstream processors: improving both performance and fault tolerance. In *Proceedings of the Symposium on Architectural Support for Programming Languages and Operating Systems*, pages 257–268, 2000. 56

M. R. Thistle and B. J. Smith. A processor architecture for Horizon. In *Proceedings of the conference on Supercomputing*, pages 35–41, 1988. 70

J. E. Thornton. *Design of a Computer: The Control Data 6600*. Scott Foresman & Co, 1970. 11, 25, 67

Marc Tremblay, Jeffrey Chan, Shailender Chaudhry, Andrew W. Conigliaro, and Shing Sheung Tse. The MAJC architecture: A synthesis of parallelism and scalability. *IEEE Micro*, 20(6):12–25, November 2000. 10, 18, 77

Nathan Tuck and Dean M. Tullsen. Initial observations of the simultaneous multithreading pentium 4 processor. In *Proceedings of the International Conference on Parallel Architectures and Compilation Techniques*, pages 26–34, 2003. 65, 76

Dean M. Tullsen and Jeffery A. Brown. Handling long-latency loads in a simultaneous multithreading processor. In *Proceedings of the International Symposium on Microarchitecture*, December 2001. 45, 63, 79

Dean M. Tullsen, Susan J. Eggers, and Henry M. Levy. Simultaneous multithreading: Maximizing on-chip parallelism. In *Proceedings of the International Symposium on Computer Architecture*, June 1995. 5, 12, 13, 33, 39, 42

Dean M. Tullsen, Susan J. Eggers, Joel S. Emer, Henry M. Levy, Jack L. Lo, and Rebecca L. Stamm. Exploiting choice: Instruction fetch and issue on an implementable simultaneous multithreading processor. In *Proceedings of the International Symposium on Computer Architecture*, May 1996. 33, 34, 37, 38, 44, 45, 72, 79

Dean M. Tullsen, Jack L. Lo, Susan J. Eggers, and Henry M. Levy. Supporting fine-grained synchronization on a simultaneous multithreading processor. In *Proceedings of the International Symposium on High-Performance Computer Architecture*, January 1999. 50

Eric Tune, Rakesh Kumar, Dean M. Tullsen, and Brad Calder. Balanced multithreading: Increasing throughput via a low cost multithreading hierarchy. In *Proceedings of the International Symposium on Microarchitecture*, pages 183–194, December 2004. 14, 64

Theo Ungerer, Borut Robič, and Jurij Šilc. A survey of processors with explicit multithreading. *ACM Computing Surveys*, 35(1):29–63, March 2003. 67

H. Vandierendonck and Andre Seznec. Fairness metrics for multi-threaded processors. *Computer Architecture Letters*, 10(1):4 –7, January 2011. 64, 65

D.W. Wall. Limits of instruction-level parallelism. In *International Conference on Architectural Support for Programming Languages and Operating Systems*, April 1991. 44

Steven Wallace and Nader Bagherzadeh. A scalable register file architecture for dynamically scheduled processors. In *Proceedings of the International Conference on Parallel Architectures and Compilation Techniques*, 1996. 47

Perry H. Wang, Hong Wang, Jamison D. Collins, Ralph M. Kling, and John P. Shen. Memory latency-tolerance approaches for Itanium processors: Out-of-order execution vs. speculative precomputation. In *Proceedings of the International Symposium on High-Performance Computer Architecture*, pages 167–176, 2002. 39

Fredrik Warg and Per Stenstrom. Dual-thread speculation: Two threads in the machine are worth eight in the bush. In *Proceedings of the International Symposium on Computer Architecture and High Performance Computing*, pages 91–98, 2006. 58

David L. Weaver and Tom Germond. *The SPARC architecture manual (version 9)*. Prentice-Hall, Inc., 1994. 18

W.-D. Weber and A. Gupta. Exploring the benefits of multiple hardware contexts in a multiprocessor architecture: preliminary results. In *Proceedings of the International Symposium on Computer Architecture*, pages 273–280, 1989. 9

Wayne Yamamoto and Mario Nemirovsky. Increasing superscalar performance through multistreaming. In *Proceedings of the International Conference on Parallel Architectures and Compilation Techniques*, pages 49–58, September 1995. 12, 33, 39

Wayne Yamamoto, Mauricio Serrano, Roger Wood, and Mario Nemirovsky. Performance estimation of multistreamed, superscalar processors. pages 195–204, January 1994. 33

Kenneth C. Yeager. The MIPS R10000 superscalar microprocessor. *IEEE Micro*, 16(2):28–40, April 1996. 34

Weifeng Zhang, Brad Calder, and Dean M. Tullsen. An event-driven multithreaded dynamic optimization framework. In *Proceedings of the International Conference on Parallel Architectures and Compilation Techniques*, pages 87–98, 2005. 54, 55

Weifeng Zhang, Dean M. Tullsen, and Brad Calder. Accelerating and adapting precomputation threads for effcient prefetching. In *Proceedings of the International Symposium on High-Performance Computer Architecture*, pages 85–95, 2007. 54, 55

Z. Zhu and Z. Zhang. A performance comparison of DRAM memory system optimizations for SMT processors. In *International Symposium on High-Performance Computer Architecture*, pages 213–224, 2005. 43

Craig Zilles and Gurindar Sohi. Execution-based prediction using speculative slices. In *Proceedings of the International Symposium on Computer Architecture*, pages 2–13, 2001. 54, 55

Zilog, editor. *Zilog Z-80 Data Book*. 1978. 17

# Authors' Biographies

## MARIO D. NEMIROVSKY

**Mario D. Nemirovsky** is an ICREA Research Professor at the Barcelona Supercomputer Center, where he has been since 2007. He holds 62 patents and has authored over 30 research papers. Mario is a pioneer in multithreaded hardware-based processor architectures. During his tenure with the University of California, Santa Barbara, Mario co-authored some of the seminal works on simultaneous multithreading. Mario has made key contributions to other areas of computer architecture, including high performance, real-time, and network processors. He founded ConSentry Networks, Inc. where he served as CTO and VP Chief Scientist. He was the architect of ConSentry high performance processor (LSP-1) in which he pioneered the concept of Massively Multithreading (MMT). Earlier, Mario founded Flowstorm and XStream Logic, Inc. Before that, he was a chief architect at National Semiconductor, PI Researcher at Apple Computers, and Chief Architect at Weitek Inc. As chief architect at Delco Electronics, General Motors (GM), he architected the GM multithread engine controller. He received his Ph.D. in ECE from University of California, Santa Barbara in 1990.

## DEAN M. TULLSEN

**Dean Tullsen** is a Professor in the Computer Science and Engineering department at the University of California, San Diego, where he has been since 1996. He has authored over 90 research papers and hold 4 patents. In addition to co-authoring some of the seminal works on simultaneous multithreading, he and his co-authors have introduced many concepts to the research community, including speculative precomputation, symbiotic job scheduling, critical-path prediction, conjoined cores, single-ISA heterogeneous multicore architectures, balanced multithreading, and data-triggered threads. He received his B.S. and M.S. in computer engineering from UCLA, and his Ph.D. from University of Washington. He is a fellow of the IEEE and a fellow of the ACM. He has twice had papers selected for the International Symposium on Computer Architecture Influential Paper Award.